Basic Life Science Methods

Basic Life Science Methods
A Laboratory Manual for Students and Researchers

Javeed Ahmad Tantray

Department of Zoology, Central University of Kashmir, Ganderbal, Jammu and Kashmir, India

Nighat Un Nissa

Department of Zoology, University of Kashmir, Srinagar, Jammu and Kashmir, India

Rasy Fayaz Choh Wani

Department of Zoology, University of Kashmir, Srinagar, Jammu and Kashmir, India

Sheikh Mansoor Shafi

Advance Centre for Human Genetics, Sheri Kashmir Institute of Medical Sciences, Srinagar, Jammu and Kashmir, India

ACADEMIC PRESS

An imprint of Elsevier

ELSEVIER

Academic Press is an imprint of Elsevier
125 London Wall, London EC2Y 5AS, United Kingdom
525 B Street, Suite 1650, San Diego, CA 92101, United States
50 Hampshire Street, 5th Floor, Cambridge, MA 02139, United States
The Boulevard, Langford Lane, Kidlington, Oxford OX5 1GB, United Kingdom

Notices
Knowledge and best practice in this field are constantly changing. As new research and experience broaden our understanding, changes in research methods, professional practices, or medical treatment may become necessary.

Practitioners and researchers must always rely on their own experience and knowledge in evaluating and using any information, methods, compounds, or experiments described herein. In using such information or methods they should be mindful of their own safety and the safety of others, including parties for whom they have a professional responsibility.

To the fullest extent of the law, neither the Publisher nor the authors, contributors, or editors, assume any liability for any injury and/or damage to persons or property as a matter of products liability, negligence or otherwise, or from any use or operation of any methods, products, instructions, or ideas contained in the material herein.

ISBN: 978-0-443-19174-9

For information on all Academic Press publications visit our website at
https://www.elsevier.com/books-and-journals

Publisher: Andre G. Wolff
Acquisitions Editor: Andre G. Wolff
Editorial Project Manager: Kristi Anderson
Production Project Manager: Kiruthika Govindaraju
Cover Designer: Mark Rogers

Typeset by TNQ Technologies

Contents

List of figures

List of tables

About the authors

Dr. Javeed Ahmad Tantray's current research focus is on human cardiovascular diseases. He obtained his Ph.D. from Osmania University in Hyderabad, India. Dr. Tantray has done his Postdoc studies with the Department of Zoology, University of Kashmir, receiving funding from the UGC-CSIR. He published many national and international peer-reviewed research articles. Dr. Tantray has published in high-impact factor journals such as the *International Journal of Cardiology* and the *Saudi Journal of Biological Sciences*. He has guided 50 M.Sc./and graduate students' projects. He has received the Junior Scientist Award both at the national and international levels. He received  Best Abstract Award at the 4th Asia-Pacific CardioMetabolic Syndrome Congress held by the Korean Society of CardioMetabolic Syndrome, with an award amount of USD 500. Dr. Tantray has organized a national seminar on "Cancer Biomarkers" and a one-day national workshop on "Assembly and Usage of Foldscope." Dr. Tantray is currently an Assistant Professor in the Department of Zoology, School of Life Sciences, Central University of Kashmir (J&K).

Nighat Un Nissa has completed her M.Sc., M.Phil., and Ph.D. from the Department of Zoology, University of Kashmir. Dr. Nissa has worked on fish parasites and their impact on pollution. She has published her research work in national and international journals. She was awarded for poster presentation at the Animal Science Congress held by the Department of Zoology, University of Kashmir.

Rasy Fayaz Choh Wani has completed her M.Sc. degree from Kashmir University and is currently pursuing her Ph.D. She has published her research work in many reputed journals. She has presented her work at different international and national conferences. She is currently working on the research topic "Isolation and characterization of pathogenic bacterial microflora associated with cultured *Cyprinus carpio* in Kashmir valley." She has delivered her services as a subject expert at EMMC, University of Kashmir, in SWAYAM PRABHA (DTH), in Zoology, launched by MHRD, Government of India.

Dr. Sheikh Mansoor Shafi is a Postdoctoral Researcher, presently working at the SK-Institute of Medical Sciences, Soura, Srinagar. Previously, he was working as Project Associate-I at the CSIR-Indian Institute of Integrative Medicine, Jammu (IIIM). He has obtained his B.Sc. from SP College of Science (University of Kashmir), M.Sc. from HNBGU, Uttarakhand, and a Ph.D. from Sher-e-Kashmir University of Agricultural Sciences and Technology of Jammu. Dr. Mansoor has made important contributions to his field and has been working hard to shape himself as an impressive and contributive scientist of the highest intellectual caliber. He has an impressive lab expe-

rience in biochemistry, molecular biology, and genetics to his credit. He has collaboratively worked and published more than 30 research and review papers in highly reputed journals like Frontiers, *PLOS One*, *Scientific Reports*, *JoF* MDPI, etc. He has published more than 10 book chapters and he has filed one patent application, which was published on 01/04/2022. Dr. Mansoor has delivered many guest lectures on sequence alignment, submission tools, and phylogenetic analysis. He has submitted more than 150 sequences to NCBI GenBank. He has attended several national and international conferences and workshops like European Molecular Biology Organization (EMBO) and also received hands-on training on Advance Techniques in Modern Biology at Jamia Hamdard University, New Delhi. Dr. Mansoor is a Review Editor in Frontiers in Ethnopharmacology and is also the reviewer of several reputed journals like 3 *Biotech*, *Cell Reports*, Frontiers, *PLOS One*, MDPI, etc.

Foreword

Life sciences have always been an important and fundamental area of science since ages. The exponential increase in the scientific information and the rate at which new discoveries are made has necessitated the consolidation of research and experimental methodologies including basic and advanced techniques. The demolition of barriers envisaged through NEP 2020 has now mandated interdisciplinary studies, and hence the importance of this book is a worth to read and consult.

This book has been designed in such a way that all the levels of the students and scholars get benefited. The basic and applied experimentations and methodologies in biology have been consolidated and expressed in a very simple and lucid way. This book shall also be useful for researchers in biology for understanding and undertaking some basic and advanced techniques.

I understand that there is always scope for improvement. The authors need to include most recent technologies and techniques for the advanced researchers to make it as a one stop manuscript.

This book is all worth to be consulted and I wish the authors great

Prof. M. Afzal Zargar
Registrar Professor in Biotechnology
Central University of Kashmir, Ganderbal
J&K India

Foreword

The science laboratory is an important resource input for teaching science and developing scientific temper. Engaging the students in science laboratory experiments enhances their basic understanding and comprehension of students. The absence of laboratory practical activities makes students and researchers lose interest in research and experimentation. In this regard, the book titled "Basic Life Science Methods—A laboratory Manual for Students and Researchers" is an incredible achievement that will cater to the basic laboratory skills required for the students and researchers. The authors have designed ready-to-use protocols that show step-by-step instructions for implementing the techniques, also approaches toward the interpretation of results have been discussed in a very lucid manner.

This book is anticipated to be an important resource for life science/biology/Agriculture undergraduates, postgraduates, and doctorate students, as well as faculties, embarking on quantitative or qualitative research projects and other experiments.

I congratulate Dr. Sheikh Mansoor and other coauthors for producing this laboratory manual, and hope it is received well by all stakeholders.

Prof. Nazir A. Ganai
Vice-Chancellor
Sher-e-Kashmir
University of Agriculture Sciences and
Technology of Kashmir
Place: Shalimar
Dated: 03-06-2022

Introduction

This one-of-a-kind, practical manual with the majority of the fundamental bioscience laboratory procedures discusses basic calculations and solution preparation, aseptic techniques, spectrophotometry, chromatography of small and big molecules; and protein and nucleic acid electrophoresis, etc. Furthermore, this book contains clear, relevant graphics and worked examples of computations. The importance of fundamental laboratory abilities is highlighted, and boxed material gives step-by-step laboratory technique instructions for ease of reference at any stage in the students' progress. Students can use this book to be updated with practical knowledge. This book explores a broad variety of modern molecular biology techniques that can be used by researchers. The authors have made the protocol ready to use by calculating the amount or the volume of the chemical or solution to be used. This book contains simple, step-by-step instructions for implementing the techniques discussed and also gives a clear-cut idea of how to interpret results from the techniques. This book (Manual) is filled with real-life examples, and figures that demonstrate the breadth of research and offer practical advice on how to improve key research skills such as critical thinking and argument construction. This practical guide focuses on assisting students, researchers, scientists, and practitioners in improving and advancing their own methodologies. Graduate students/researchers of streams like Biotechnology, Microbiology, Biochemistry, Zoology, Botany, Environmental Science, Agriculture, etc., will be benefited from this manual. In summary, this is a "must-have" for any early life science/agriculture students who are failing to grasp this critical component of their course.

Calculations of molarity and normality

Chapter outline

Molarity (M)

Molarity of a solution is the number of moles of the solute per liters of solution (or number of millimoles per mL of solution). Unit of molarity is mol/L or mol/dm^3 for example; a molar (1M) solution of sugar means a solution containing 1 mol of sugar (i.e., 342 g or 6.02×10^{23} molecules of it) per liter of the solution. Solutions in term of molarity generally expressed as

➢ 1M = Molar solution, 2M = Molarity is two,
➢ M/2 or 0.5 M = Semimolar solution,
➢ M/10 or 0.1 M = Decimolar solution
➢ M/100 or 0.01 M = Centimolar solution
➢ M/1000 or 0.001 M = Millimolar solution

Molarity is most common way of representing the concentration of solution. It depends on temperature as $M \propto 1/T$.

When a solution is diluted ($\times$ times), its molarity also decreases (by $\times$ times). Mathematically molarity can be calculated by following formulas:

➢ M = No. of moles of solute (n)/Vol. of solution in liters
➢ M = Wt. of solute (in gm) per liter of solution/Mol. wt. of solute
➢ M = Wt. of solute (in gm)/Mol. wt. of solute $\times$ 1000/Vol. of solution in mL.
➢ M = No. of millimoles of solute/Vol. of solution in mL
➢ M = Percent of solute $\times$ 10/Mol. wt. of solute
➢ M = Strength in gL/1 of solution/Mol. wt. of solute

Basic Life Science Methods. https://doi.org/10.1016/B978-0-443-19174-9.00001-5

- ➤ $M = 10 \times$ Sp. gr. of the solution $\times$ Wt. % of the solute/Mol. wt. of the solute
- ➤ If molarity and volume of solution are changed from M_1, V_1 to M_2, V_2. Then,

$$M_1 V_1 = M_2 V_2 \text{(Molarity equation)}$$

- ➤ In balanced chemical equation, if n_1 moles of reactant one reacts with n_2 moles of reactant two, then

$$M_1 V_1/n_1 = M_2 V_2/n_2$$

- ➤ If two solutions of the same solute are mixed, then molarity (M) of resulting solution is as follows:

$$M = M_1 V_1 + M_2 V_2/(V_1 + V_2)$$

- ➤ Volume of water added to get a solution of molarity M_2 from V_1 mL of molarity M_1 is

$$V_2 - V_1 = (M_1 - M_2 / M_2)V_1$$

Normality (N)

Normality is a measure of concentration equal to the gram equivalent weight per liter of solution. Gram equivalent weight is the measure of the reactive capacity of a molecule. The solution's role in the reaction determines the solution's normality.

For acid reactions, a 1M H_2SO_4 solution will have normality (N) of 2 N because 2 mol of H^+ ions are present per liter of solution.

For sulfide precipitation reactions, where the SO^- ion is the important part, the same 1M H_2SO_4 solution will have a normality of 1 N.

Relation between molarity and normality

Normality of solution = Molarity $\times$ Molecular mass/Equivalent mass
Normality $\times$ Equivalent mass = Molarity $\times$ Molecular mass.
For an acid, Molecular mass/Equivalent mass = Basicity So,
Normality of acid = Molarity $\times$ Basicity.
For a base, Molecular mass/Equivalent mass = Acidity So,
Normality of base = Molarity $\times$ Acidity.

Percentage solution

A relationship of a quantity of solute to the quantity of solution, multiplied by 100, expressed in terms of mass of solute per mass of solution. It can be expressed in terms of mass solute per mass solution, volume solute per volume solution, or mass solute (g) per volume (mL) solution. An example of a 5% by mass solution is 5 g of glucose dissolved in 95 g of water, forming 100 g of solution.

Observations and result

1. Molarity = Amount of solute × 1000/Molecular mass × volume (mL)
   E.g., Find the amount of solute if 5M NaCl was dissolved in 30 mL.
   5M = Amount of solute × 1000/58.5 × 30
   Amount of solute = 30 × 58.5 × 5/1000
   Amount of solute = 8.775 gm

2. Normality = Amount of solute × 1000/Equivalent mass × volume (mL)
   [Equivalent mass = Molecular mass/valency]
   E.g., Find the amount of solute if N/10 NaOH was dissolved in 30 mL.
   1/10 = Amount of solute × 1000/1 × 30
   Amount of solute = 1200/1000 × 10
   Amount of solute = 0.12 gm

3. Percentage = part/whole = percent/100
   E.g., Find the amount of acetone and water if 10% acetone dissolved in 1 mL.
   10 × 1/100 = 0.1
   Amount of acetone = 0.1 mL
   Amount of water = total solution − amount of acetone in solution
   − 10 ▬ 0.1
   = 9.9 mL

4. Parts per million = part/whole = ppm/10^6
   E.g., A ceplox tablet of 50 mg what is the amount of water needed to prepare of 1000 ppm?
   50 × 10^{-3}/whole = 1000/10^6
   = 50 × 10^{-3} × 10^6/1000
   = 50 mL

5. Dilution = $N_1V_1 = N_2V_2$
   E.g., What amount we take from 1000 ppm to prepare 10 mL of 60 ppm, and how much distilled water is added?
   1000 × V_1 = 60 × 10
   V_1 = 60 × 10/1000
   = 0.6 mL from 1000 ppm
   Amount of water = total amount − V_1
   = 10 − 0.6
   = 9.4 mL of distilled water is added.

Glassware washing and sterilization techniques

Chapter outline

Principle

Cleaning of Glasswares by chromic acid

Cleaning of glasswares is a very important step in every laboratory as it ensures a clean and sterile environment to work with. The general mechanisms for cleaning of various types of glasswares can be summarized as follows:

Washing of glasswares is done by two main methods

1. Cleaning by detergents
2. Cleaning by chromic acid

General procedure for cleaning of glasswares by detergent

➢ Choose unclean glasswares
➢ Wash it with tap water (if still feels greasy/unclean)
➢ Wash with detergent solution
➢ Rinse with tap water till it fully gets clean
➢ Glass wares are air dried

    (Note: Keep only ungraduated glass in oven for drying).

Basic Life Science Methods. https://doi.org/10.1016/B978-0-443-19174-9.00002-7

General procedure for cleaning of glasswares by chromic acid

Chromic acid is a strong oxidizing agent that can very well remove all the impurities present in form of adhering minerals, organic impurities, and carbonates present in the glassware, which cannot be washed out through general washing.

Preparation of 5% chromic acid

(1) Take a glass bowl filled with ice flakes
(2) Keep a 500 mL conical flask put into bowl (Because when we add conc. H_2SO_4 in $K_2Cr_2O_7$ it produces heat. It is an exothermic reaction.)
(3) Put 5 g $K_2Cr_2O_7$ in the conical flask and add 10 mL distilled water
(4) Add 100 ml conc. H_2SO_4 slowly with constant shaking

Cleaning by chromic acid

(1) Take unclean glass wares.
(2) Rinse in chromic acid and leave for 4–5 h in a closed vessel.
(3) Wash with running tap water.
(4) Rinse with distilled water and keep them in oven for drying.

Types of sterilization

➤ **Flame sterilization**: The inoculation loop, scalpel can be sterilized by dipping in 70% alcohol and then heating by flame.
➤ **Sterilization by autoclave**: By autoclaving glasswares like flasks, test tubes, petridishes, and media can be autoclaved at 121°C, 15 Ibs pressure, for 15 min.
➤ **Sterilization by filtration**: Some chemicals and vitamins are used. They are temperature sensitive. Due to autoclaving, their protein structure will get denatured, that is why they are sterilized by filter that has porosity 0.22 μ.
➤ **Dry hot air sterilization**: Hot-air ovens are most commonly used for sterilizing glassware like petridish, test tubes, pipettes metal instruments that can tolerate prolonged heat exposure, oil, powders, waxes, and other articles that are either solid or not effectively sterilized by the moist heat of the autoclave. Sterilization is accomplished by exposure of items to 150–180°C.

Observation and results

5% chromic acid solution is prepared by mixing 1 g of potassium dichromate in 5 mL of distilled water and 15 mL of sulfuric acid and the solution is used to clean the unclean glassware. These glasswares are kept for further use in laboratory.

pH meter: its use and calibration

3

Chapter outline

Aim: To adjust/check the pH of any given solution.

Structure

A general pH meter contains one pH probe (it is a glass rod, with a bulb like structure at its lower end and connected to electric wire at another end). Some pH meter also contains a second probe for temperature calibration, because pH changes with change in temperature, otherwise in modern pH meters both pH probe and temperature probe are united in a single probe.

✔ NOTE: The Probe (Glass Rod Like) Must Never Be Allowed to Dry in Air. Always Must Be Kept Dipped In 3 m KCl Solution.

Calibration

1. Switch on the power of pH meter.
2. Take the pH probe out of KCl solution and wash it properly with distilled water.
3. Now dip the pH probe inside a standard buffer of pH 7 and calibrate it by pressing calibration button on the system when it displays its pH.
4. Calibration is successfully complete.

Basic Life Science Methods. https://doi.org/10.1016/B978-0-443-19174-9.00003-9

Adjusting/Maintaining the pH of a solution:

1. Switch on the power of pH meter.
2. Take the pH meter probe out of KCl solution and wash properly with distilled water.
3. Dip the probe in the solution, of which pH is to be adjusted and see the reading of pH on its display.
4. Wait for 1—2 min till the reading is stabilized.
5. This reading is the actual pH of that solution.
6. If you want to set the pH of your solution toward basicity, then add 10 N NaOH solution or 20% HCl solution if you want to set the pH of your solution toward acidity with the instruction given below:
   - ✔ **Always add acid or base slowly and drop by drop to your solution and keep on shaking and mixing the solution by hand in between.**
   - ✔ **Never use concentrated acid for adjusting the pH.**

Observation and result

E.g., for preparing 0.1 M of sodium phosphate monobasic solution pH 6.2 with the help of pH meter (Fig. 3.1).

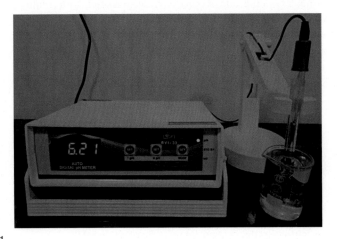

FIGURE 3.1

pH meter machine.

Principle and working of autoclave

Chapter outline

Principle: An autoclave is a device that sterilizes equipment using the moist heat sterilization concept, in which saturated steam is created under pressure to kill microorganisms such as bacteria, viruses, and even heat-resistant endospores. This is accomplished by raising the temperature of the instruments within the device above the boiling point of water. Gas laws, which indicate that the higher the pressure within the device, the higher the temperature rises, are likewise included in this process. To put it another way, pressure and temperature are exactly proportional (Figs. 4.1 and 4.2).

Autoclaves normally achieve a temperature of around 121°C, and the sterilization procedure takes around 15–20 min. Autoclave cycles, on the other hand, may be regulated by the working technician. Solids, liquids, hollows, and other tools of various forms and sizes may all be sterilized in autoclaves. Surgical equipment, pharmaceutical goods, laboratory tools, and a variety of other items are instances of this. Culture media, autoclavable plastic materials, solutions and water, selective glassware, pipette tips, plastic tubes, and biohazardous waste are all examples of what autoclaves can sterilize.

Different temperature ranges are used for steam sterilization such as 250°F (121°C), 270°F (132°C), or 275°F (135°C). But, 121°C (250°F) and 132°C (270°F) are the two most common steam-sterilizing temperatures used to kill microorganisms.

Basic Life Science Methods. https://doi.org/10.1016/B978-0-443-19174-9.00004-0

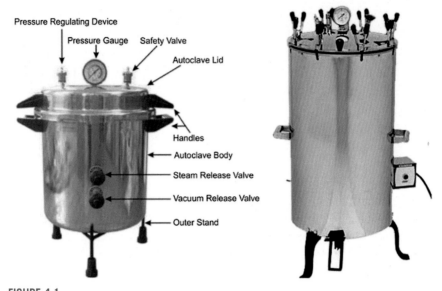

FIGURE 4.1
Representative pictures of autoclave.

Conditions for autoclave sterilization

Organism	Vegetative cells	Spores
Yeasts	5 min at 50–60°C	5 min at 70–80°C
Molds	30 min at 62°C	30 min at 80°C
Bacteria	10 min at 60–70°C	2 to over 800 min at °C 0.5–12 min at 121°C
Viruses	30 min at 60°C	

(1) Pressure Cooker Type/Laboratory Bench Autoclaves: This type of autoclave is widely utilized all over the world. It has an air and steam discharge tap, as well as a pressure gauge and a safety valve. At the bottom of the chamber, there is also an electric immersion heater.

(2) Gravity Displacement Autoclave: In laboratories, they are widely used. This autoclave uses a heating unit to create steam within the chamber, which may circulate around the chamber for efficient sterilization. In comparison to other autoclaves, it is also quite inexpensive.

(3) Positive Pressure Displacement Autoclave (B-type): This type of autoclave uses a separate steam generator unit to generate steam, which is then transferred into the autoclave. Steam may be created in seconds; hence it is known to be faster.

(4) Negative Pressure Displacement Autoclave (S-type): A steam generator and a vacuum generator are both included in negative pressure displacement autoclaves. The vacuum generator draws all of the air out of the autoclave, whereas the steam generator, like the positive pressure displacement autoclave, generates heat and sends it into it.

Procedure for operating an autoclave

i. Check to see if the chamber has any prior instruments.
ii. Fill the chamber with water and double-check the quantity.
iii. Inside the chamber, place the instruments.
iv. Switch on the electric heater after closing the lid and tightening the screws.
v. Maintain the proper pressure level within the chamber by adjusting the safety valves.
vi. When the water within the chamber boils, the air—water combination can exit through the discharge tube, displacing all of the air inside. When no more water bubbles emerge from the pipe, the displacement is complete.
vii. Allow the steam to reach the required level by closing the drainage pipe.
viii. Blow the whistle once the pressure level has been achieved to release all of the surplus pressure in the chamber and after the whistle, let the autoclave run for the specified amount of time.
ix. Turn off the electric heater and let the autoclave cool until the pressure within the chamber reaches atmospheric levels.
x. Allow air from outside the autoclave to enter by opening the discharge line. Remove the instruments from the compartment by opening the lid.

Autoclave pressure and temperature chart

Sterilizer	Temperature	Pressure	Time
Steam autoclave	121°C	15 psi	15 min
Unwrapped items	132°C	30 psi	3 min
Dry heat wrapped	170°C		60 min
	160°C		120 min
	150°C		150 min
	140°C		180 min
	121°C		12 h
Dry heat (rapid flow) unwrapped items	190°C		6 min
Ethylene oxide	Ambient		8—10 h

Application of autoclave

- It is used to sterilize glassware, tools, and media in labs.
- Medical equipment, glassware, surgical equipment, and medical wastes are sterilized in autoclaves at medical laboratories.
- Medical waste including germs, viruses, and other biological contaminants can also be decontaminated using an autoclave.

Principle and working of centrifuge

Chapter outline

Centrifugation is a process that uses centrifugal force to separate mixtures. A centrifuge is a device that, often powered by an electric motor, rotates an item, such as a rotor, around a fixed axis.

Principle: The centrifuge operates on the concept of gravity and the development of centrifugal force to sediment distinct fractions. The rate of sedimentation is determined by the applied centrifugal field (G) being directed radially outward. G is determined by

- Angular velocity (ω in radians/sec)
- Radial distance (r in cm) of the particle from the axis of rotation
- $G = \omega^2 r$

Rate of sedimentation

It is determined by elements other than centrifugal force, such as particle mass, density and volume, medium density, particle form, and friction.

Basic Life Science Methods. https://doi.org/10.1016/B978-0-443-19174-9.00005-2

Sedimentation time

It depends on the following factors and they are

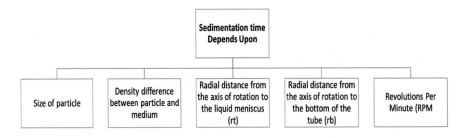

Parts of centrifuge

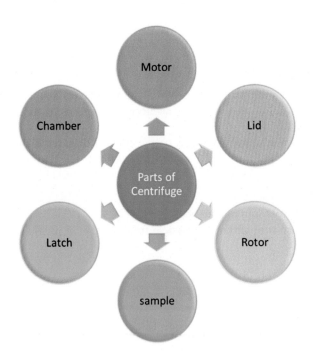

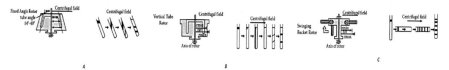

FIGURE 5.1

Three different kinds of rotors: (A) fixed angle rotors, (B) vertical rotors, and (C) swinging bucket rotors/horizontal rotors.

- **Rotor**: The motor devices that contain the tubes holding the samples in centrifuges are known as rotors. Centrifuge rotors are intended to provide the rotational speed that allows components in a sample to separate. There are three types of rotors fixed-angle rotors, vertical rotors, and swinging bucket rotors/horizontal rotors.

    i. **Fixed angle rotors**—Tubes are held at 14—40 degrees to the vertical. Particles move in a radial direction and travel a short distance. Effective for differential centrifugation. Reorientation of the tube when the rotor accelerates and decelerates (Fig. 5.1A).

    ii. **Vertical rotors**—Parallel to the rotor axis, held vertically. Particles travel over small distances. The separation time is shorter. The disadvantage of this is that the pellet may fall back into the solution at the end of the centrifugation process (Fig. 5.1B).

    iii. **Swinging bucket rotors/horizontal rotors**—When the rotor accelerate sign out to a horizontal position. A greater distance of travel, as in density gradient centrifugation, may allow for better separation. It is easier to remove the supernatant without damaging the particle. Typically employed in density-gradient centrifugation (Fig. 5.1C).

Relative centrifugal force

Relative centrifugal force (RCF) is a measure of the strength of various types and sizes of rotors. This is the force exerted on the rotor's contents as a result of the revolution. The perpendicular force operating on the sample that is constantly related to the gravity of the earth is referred to as relative centrifugal force. The RCF of each centrifuge may be used to compare rotors, allowing the best centrifuge for a specific function to be chosen. The RCF of a centrifuge may be computed using the following formula, which uses revolutions per minute (RPM) and the radius of the rotor (r). The formula for calculating the RCF is as follows:

$$\text{RCF(g Force)} = 1.118 \times 10^{-5} \times r \times (\text{RPM})^2$$

where r is the radius of the rotor (in centimetres), and RPM is the speed of the rotor in revolutions per minute.

Types of centrifuges

Centrifuges vary in general design and size depending on their intended function. The central motor that spins a rotor carrying the materials to be separated is a characteristic shared by all centrifuges (Fig. 5.2). Centrifuge types are determined by factors such as maximum sedimentation speed, vacuum presence or absence, temperature control refrigeration, sample volume and capacity of centrifugation tubes, and so on.

1. Small Benchtop Centrifuge

   It is compact and basic, with a maximum speed of 3000 rpm. It lacks a temperature control mechanism and is often used to collect fast sedimenting material such as blood cells, yeast cells, or bulky precipitates of chemical reactions, which are prevalent in the clinical laboratory for blood plasma or serum separation, urine, and bodily fluids separation. Depending on the diameter, it can take around (up to) 100 tubes.

2. Microcentrifuge

   Microcentrifuges, also known as microfuges or Eppendorf centrifuges, use small-volume tubes (up to 2 mL i.e., microcentrifuge tubes). It is widely used in medical laboratory sections such as biochemistry, molecular biology, microbiology, immunology/serology, blood banking, and laboratory pharmaceuticals, among others. With or without refrigeration, it can create forces of up to $15,000 \times g$.

3. High-Speed Centrifuges

   The highest speed is 25,000 rpm, and the centrifugal force is 90,000 g. Refrigeration is installed to remove the heat generated. The thermocouple had been used to keep the temperature constant at 0–4°C. It is used to collect bacteria, cell debris, cells, large cellular organelles, chemical reaction precipitates, and it is also effective in separating subcellular organelles such as nuclei, mitochondria, and lysosomes.

4. Ultra-Centrifuge

   It has the following characteristics:

   Necessitate customized rotors, Because the high speeds employed in such instruments make a significant quantity of heat, cooling mechanisms, and vacuum are necessary in ultracentrifuges. At a speed of 75,000 rpm, it generates a

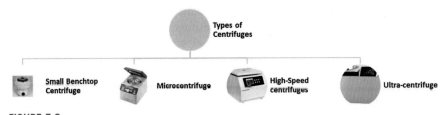

FIGURE 5.2

Different types of centrifuges.

centrifugal force of 500,000 g. To achieve a vacuum, the pump locks and evacuates the rotor chamber. Refrigeration system (temperature range: 0−4°C). The rotor chamber is usually surrounded by a thick Armor plate.

Type of centrifugations
Centrifugation

Centrifugation for isolation and purification of components is known as preparatory centrifugation, while that carried out with a desire for characterization is known as analytical centrifugation.

Preparative centrifugation

It is focused with the actual isolation of biological material for future biochemical analyses and is classified into two primary approaches based on the suspension medium in which separation occurs

- Homogenous medium—differential centrifugation
- Density gradient medium—density gradient centrifugation
    a. **Differential centrifugation**: Differential centrifugation separates particles based on particle size. It is often utilized in basic pelleting and the partial purification of subcellular organelles and macromolecules. It is used to investigate subcellular organelles, tissues, or cells (first disrupted to study internal content). Larger particles settle quicker than smaller particles during centrifugation. A succession of increasing g-forces results in largely filtered organelles. Despite its lower yield, differential centrifugation is still the most often used technique for isolating intracellular organelles from tissue homogenates due to its ease of use. It is convenient and saves time, but the negative is the low yield and the fact that the preparation obtained is never pure.
    b. **Density gradient centrifugation**: It is the technique of choice for purifying subcellular organelles and macromolecules. A density gradient can be created by layering gradient media, such as sucrose, in a tube, with the heaviest layer at the bottom and the lightest at the top. There are two types of classifications:
- Rate-zonal (size) separation
- Isopycnic (density) separation
    Gradient materials used are Sucrose (66%, 50°C), Silica solution—Glycerol, CsCl, Cs Acetate, ficol, sorbitol, polyvinylpyrrolidone.
    i. **Rate zonal centrifugation**: It is gradient centrifugation, and it uses particle size and mass rather than particle density for sedimentation. Separation of biological organelles, such as endosomes or proteins, is a popular use (such as antibodies).

ii. **Isopycnic centrifugation**: During centrifugation, a particle of a specific density will sink until it reaches a place where the density of the surrounding solution is precisely the same as the particle's density. Once quasi-equilibrium is attained, the length of centrifugation has little effect on particle movement. For example, nucleic acid separation in a CsCl (cesium chloride) gradient.

Analytical centrifugation

It has a maximum speed of 70,000 rpm and an RCF of up to 500,000 g. Motor, rotor, chilled and evacuated chamber, optical system. The optical system consists of 2 cells—an analytical cell and a counterpoise cell—as well as a light absorption system, a schlieren system, and a Rayleigh interferometric system. Schlieren optics or Rayleigh interference optics were utilized. The peak of the refractive index will be at the meniscus at first. As sedimentation progresses, macromolecules migrate downward and the peak moves, providing immediate information about the sedimentation properties. It is used to examine macromolecule purity, relative molecular mass of solute (within 5% SD), change in relative molecular mass of super molecular complexes, protein structure conformational change, and ligand-binding studies.

Uses of centrifuge

The centrifuge is used in the clinical laboratory for the following purposes:

1. Separate cellular materials from blood in order to obtain cell-free plasma or serum for examination.
2. Clean an analytical specimen of chemically precipitated protein.
3. Separate protein-bound ligand from free ligand using immunochemical and other methods.
4. Subcellular organelle, DNA, and RNA separation.
5. Convert aqueous to organic solvents to extract solutes in biological fluids.

Principle and working of laminar

Chapter outline

Laminar flow is defined as airflow in which the velocity and direction of the entire body of air inside a given space are the same. In microbiology laboratories, a laminar air flow is a frequent piece of equipment. It consists of a chamber with an air blower attached to the back side that allows air to flow in straight lines that are parallel to each other at a consistent rate. A laminar flow cabinet/main hood's purpose is to produce a contaminant-free working environment. It filters and captures all types of impurity particles that enter the cabinet. To eliminate airborne impurity particles as small as 0.3 μm, it uses a filter pad and a particular filter system known as a high-efficiency particulate air filter, or HEPA filter. A tissue culture hood or a laminar flow closet are other names for a laminar air flow chamber.

Principle: The laminar air flow chamber works on the principle of laminar air flow, as the name suggests. The flow of air is said to be laminar when the gas molecules travel in many straight lines that are parallel to each other. The gas molecules do not collide with one another as they travel. It uses a high-efficiency particulate airflow technology that catches and eliminates all forms of airborne impure particles to keep the environment clean and sterile.

Parts of a laminar air flow chamber (Fig. 6.1)

1. **Cabinet**: Stainless steel is commonly used for the cabinet of a laminar air flow chamber. It protects the sterile environment established within the laminar air flow hood from contamination and impure particles found outside by providing

Basic Life Science Methods. https://doi.org/10.1016/B978-0-443-19174-9.00006-4

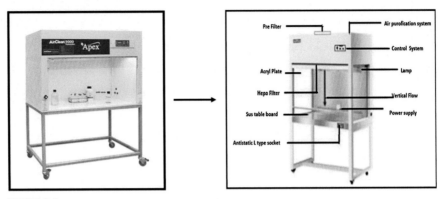

FIGURE 6.1

Laminar air flow chamber.

insulation. A glass shield on the front of the cabinet allows the user to access the cabinet by offering a partial or whole aperture.

2. **Working Station**: Inside the chamber, a flat working station offers a stable platform for activities such as plant tissue culture, electronic wafer creation, microorganism growth, and so on. It aids with the placement of culture plates, burners, samples, and other instruments.

3. **Filter Pad**: The filter pad is used to block or capture the impure particles and prohibit them from being transmitted any further. The filter pad is also known as the prefilter or the primary filter as it initially sucks the air and performs the first stage filtering of air. In a vertical laminar air flow cabinet, a filter pad is placed on the top of the device, whereas in the case of a horizontal laminar air flow cabinet, it is fixed at the bottom of the chamber. The impure particles of size 5 μm or higher typically get trapped by a filter pad.

4. **Fan or Blower**: The fan or the blower sucks the prefiltered air through the filter pad and transmits it toward the high-efficiency particulate air filter. In a vertical laminar air flow cabinet, the blower is usually present right below the filter pad. On the contrary, the position of the fan or blower in the case of a horizontal laminar air flow cabinet is right next to the filter pad.

5. **HEPA Filter**: HEPA filter or high-efficiency particulate air filter is a special air filter present inside the chamber that helps in the removal of all sorts of contamination particles including bacteria, fungi, and dust particles to maintain a safe and sterile environment. For this purpose, the prefiltered air is made to pass through the HEPA filter, which acts as the secondary or final filter. The particles that are even 0.3 μm in size can be successfully eliminated with the help of a HEPA filter. To remove the impure particles, a HEPA filter generally makes use of three mechanisms as given below:

   a. **Interception**: Under this mechanism, the impure particles get stick to the filter fibers. The interception mechanism is used by the HEPA filter to filter out large impure particles.

b. Impaction: It involves a sudden change in the airflow that causes the particles to get embedded in the filter fibers. The impaction mechanism is also used to remove comparatively large impure particles from the inner environment of the chamber.

c. Diffusion: Here, the impure particles tend to interact with each other, move in a zig-zag path, and display Brownian motion. This random and repeated motion of particles causes them to get trapped within the filter fibers. Relatively small contaminants can be eliminated with the help of the diffusion process.

6. UV Lamp: To prevent contamination of the experiment, laminar flow cabinets may incorporate a UV-C light to sanitize the inside and contents before use. Before using the cabinet, UV lights are normally left on for 15 min to sanitize the interior.

7. Fluorescent Lamp: A fluorescent light is installed inside the cabinet to give adequate illumination throughout the procedure.

➤ Working of a Laminar Air Flow Chamber

The blower and fluorescence light are engaged when the gadget is turned on. The air is drawn in by the blower, the large impure particles are filtered out by the filter pad, and the minute impurities are removed by the HEPA filter. The device's lid is partially or entirely opened after some time. With the aid of alcohol and a clean fabric cloth or cotton, the workstation is then completely and correctly cleansed. The alcohol used for this function might be anywhere between 60% and 95% pure. The device's lid is then closed, the blower is switched off, and the UV light is turned on. UV radiation is carcinogenic, inducing alterations in the body and, as a result, cancer. As a result, the user should not be exposed to UV light for an extended period of time. The bacteria, pathogens, and other microlevel contaminants are killed by turning on the UV light for at least 15 min. After that, the UV light lamp is switched off, and the equipment is disinfected completely.

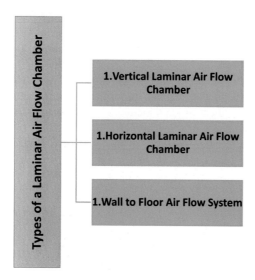

Types of a Laminar Air Flow Chamber

1. Vertical Laminar Air Flow Chamber

1. Horizontal Laminar Air Flow Chamber

1. Wall to Floor Air Flow System

Advantages of a laminar air flow chamber

- The advantages of laminar air flow devices are that they do not emit any poisonous gas into the environment.
- Laminar air flow chambers do not need to be serviced or repaired on a regular basis. As a result, they are relatively affordable and cost-effective.
- They are easily transportable and may be moved to other areas.
- During an experiment, laminar air flow cabinets limit the risks of turbulence in the surroundings.
- Some laminar air flow devices come with sophisticated security systems that will raise alerts and warn the user if there is a security breach.

Disadvantages of a laminar air flow chamber

- Placing items or hands on the device disturbs air flow, creates turbulence, and inhibits the device's capacity to disinfect the inside environment adequately.
- Some laminar air flow devices spew gases directly into the users' faces.
- For optimal performance, laminar air flow devices must be handled and cared for properly.

Applications of a laminar air flow chamber

A laminar air flow chamber may be found in a range of industries, including medical, biology laboratories, chemical industries, manufacturing plants, pharmaceutical companies, and many more. The following are some of the most common applications for a laminar air flow cabinet:

- In laboratories, a laminar air flow cabinet is used to create a sterile atmosphere for activities like plant tissue culture. This is because the presence of pollutants in the environment may readily impact these processes.
- Inside the air flow chambers, various particle-sensitive electron devices are manufactured and operated.
- The pharmaceutical business is one of the most common uses for a laminar air flow chamber. A clean and hygienic atmosphere is required for the manufacturing of medications and medicines. As a result, such procedures take place within the laminar air flow cabinets.
- Inside the laminar air flow devices, many laboratory processes such as media plate preparation, microbe culture, and so on are carried out.

When employing a laminar air flow chamber, there are a few things to keep in mind

While accessing or operating on a laminar air flow chamber, certain precautions must be observed. The following are some of the precautions and safety procedures that the user should take:

- When using the gadget, the operator must wear safety goggles, long gloves, and a lab coat.
- Before and after usage, all of the components and gadgets in the cabinet must be sanitized.
- UV light and airflow should not be utilized at the same time.
- Before and after usage, the laminar air flow cabinet must be fully irradiated with UV radiation. This helps to prevent germs and other particles from growing inside the device, creating a sterile environment.
- While the UV light is still on, any continuing procedure should be stopped immediately.

Estimation of total cholesterol (Flegg, 1972)

Chapter outline

Principle: Cholesterol esterase hydrolyzes cholesterol ester. Free cholesterol is oxidized by the cholesterol oxidase (CHO) to cholest-4en-3-one and hydrogen peroxide. Hydrogen peroxide formed reacts with four amino antipyrine and phenol in the presence of peroxide to form a dye. The density of dye formed is equal to cholesterol concentration in sample.

$$\text{Cholesterol ester} + H_2O \xrightarrow{\text{CHE}} \text{Cholesterol} + \text{fatty acid}$$

$$\text{Cholesterol} + O_2 \xrightarrow{\text{CHO}} \text{Cholest-4-en-3-one} + H_2O_2$$

$$H_2O_2 + 4AAP + O_2 \xrightarrow{\text{POD}} \text{Quinineimine (colored dye)} + H_2O_2$$

Where CHE is cholesterol esterase, CHO is cholesterol oxidase, POD is peroxidase, and AAP is amino antipyrine.

Materials and equipment (materials and reagents)

UV spectrophotometer

Cuvettes

Pipettes

37°C incubator

Enzyme reagent: A solution containing pipes buffer 35 mmol (pH 6.5 at 25°C), at least cholesterol esterase 2000 U/L, cholesterol esterase 200 U/L, cholesterol oxidase 200U/L, peroxidase 1000U/L, chromogen, surfactant, and a preservative.

Standard: Cholesterol concentration of 200 mg/dL.

Distilled water and serum sample.

Reagent Preparation:

Buffer solution is brought to room temperature and reconstituted one vail of enzyme reagent with 20 mL buffer solution.

Step-by-step method details (experimental procedure)
Procedure

All the above reagents should be pipetted into clean dry test tubes labeled Blank (B), Standard (S), and Test (T).

	(B)	(S)	(T)
Enzyme reagent	1.0 mL	1.0 mL	1.0 mL
Distilled water	10 ul	—	—
Standard	—	10 ul	—
Serum sample	—	—	10 ul

The above constituents are mixed well and incubated at 37°C for 5 min. The O.D readings are taken in a spectrophotometer at 505 nm (Hg 546). The cholesterol content is calculated as

$$\text{Cholesterol. in mg}\% = \frac{\text{Absorbance of T}}{\text{Absorbance of S}} \times \text{Standard value}$$

where A = absorbance, T = test sample, S = standard.

The cholesterol levels were expressed as mg/dL.

Expected outcomes

Total cholesterol
 Desirable <200 mg/dL (5.2 mmol/L).
 Borderline high 200–239 mg/dL (5.2–6.2 mmol/L).
 High >240 mg/dL (6.2 mmol/L).
 Each laboratory should investigate the transferability of the expected values to its own patient population and if necessary, determine its own reference range. For diagnostic purposes, the cholesterol results should always be assessed in conjunction with the patient's medical history, clinical examination, and other findings.

Estimation of HDL—cholesterol (Demacker et al., 1980)

Principle: The VLDL and LDL fractions of serum sample are precipitated using PFA and then HDL in the supernatant is separated by centrifugation and measured for its cholesterol content. The enzyme cholesterol ester hydrolase (CHE) hydrolyzes the ester cholesterol, then cholesterol is oxidized by cholesterol oxidase (CHO) to Cholest-4 en-3—one and hydrogen peroxide. Hydrogen peroxidase in the presence of enzyme peroxidase (POD) reacts with four amino antipyrine and phenol to produce red color complex, whose absorbance is proportional to HDL cholesterol concentration.

Serum + Precipitating reagent $\longrightarrow$ Precipitate (VLDL & LDL) + supernatant

Cholesterol sterol + H_2O $\xrightarrow{\text{CHE}}$ Cholesterol + fatty Acid

Cholesterol + O_2 $\xrightarrow{\text{CHO}}$ Cholest – 4–en-3- one + H_2O_2

H_2O_2 + 4 Amino anti pyrene + Phenol $\xrightarrow{\text{POD}}$ Quinoneimine dye + H_2O_2

Materials and equipment (materials and reagents)

- 10 × 75 mm test tubes
- Serum samples
- Cholesterol reagent
- Pipets
- Spectrophotometer
- Controls
- Refrigerated centrifuge
- Precipitating reagent

- Precipitating reagent, enzymatic reagent, buffer solution, HDL cholesterol standard
- Reagent preparation:

  Reconstitute one vail of enzyme reagents vail with 10 mL of buffer solution.

Step-by-step method details (experimental procedure)
It involves two steps

Step 1—Precipitation of VLDL andand LDL.
The below constituents are pipetted into a clean dry centrifuge tube.

Serum	−0.1 mL
Precipitating reagent	−0.1 mL

The constituents are mixed well and allowed to stand at room temperature for 5 min. The solution is centrifuged at 2000−3000 rpm for 10 min to get a clear supernatant.

    Step 2—Assay of HDL cholesterol.

The following constituents are pipetted into a clean dry test tubes labeled blank (B), standard test (T).

	(B)	(S)	(T)
Enzyme reagent	1.0 mL	1.0 mL	1.0 mL
Distilled water	50 ul	–	–
Serum sample	–	–	50 ul
Standard	–	50 ul	–

All the above reagents are mixed well and incubated at 37°C for 5 min. Absorbance of the test (T) and standard (S) against blank (B) is measured on spectrophotometer at 505 nm (H g 546 nm) by following formula:

$$\text{HDL Cholesterol. in mg \%} = \frac{A \text{ of}(T)}{A \text{ of}(S)} \times 100$$

where A = absorbance, T = test, S = standard.
    The HDL cholesterol is expressed by mg/dL.

Expected outcomes

The critical call results ("Panic Values") as established by NCHS is Cholesterol >400 mg/dL There is no critical call range for HDL cholesterol.

Estimation of LDL cholesterol

LDL cholesterol can be calculated by using the Friedewald's formula. LDL cholesterol mg % = total cholesterol—(HDL cholesterol + VLDL cholesterol).

Estimation of VLDL cholesterol

VLDL cholesterol can be calculated by using following formula:

$$VLDL \text{ cholesterol mg \%} = Triglycerides \times 1/5$$

Estimation of triglycerides (Fossati and Lorenzo, 1982).

Principle: Triglycerides in the sample are hydrolyzed by lipoprotein lipase to glycerol and free fatty acid. Glycerol is phosphorylated by ATP (Adenosine -5 triphosphate) to glycerol-3-phosphate in the reaction catalyzed by glycerol kinase (GK). Glycerol-3-phosphate is oxidized to dihydroxyacetone phosphate in a reaction catalyzed by enzyme glycerol phosphate oxidase (GPO). In the reaction hydrogen peroxide is produced in equimolecular concentration to the level of triglycerides present in the sample. H_2O_2 reacts with 4-amino antipyrine (4APP) and 3,5 Di choloro-2-hydroxy benzene sulfonic acid (DHBS) in a reaction catalyzed by peroxidase (POD). The result of this oxidative coupling is a quinoneimine red-colored dye. The absorbance of this dye in the solution is proportional to the concentration of triglycerides in the sample. The series of reactions involved in the assay is reported below.

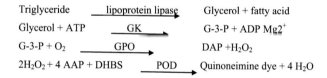

where GK is glycerol kinase, GPO is glycerol phosphate oxidase, and POD is a peroxidase.

Materials and equipment (materials and reagents)

- Triglyceride reagent
- Patient samples
- Standard
- Control serum
- UV spectrophotometer
- Cuvettes
- Pipettes
- 37°C incubator
- Working reagent: A buffered solution containing 0.4 mmol/L4-amino antipyrine, 2.6 mmol/L adenosine triphosphate, 1.0 mmol/L, ˃2400 U/L glycerol phosphate oxidase, ˃1000 U/L lipoprotein lipase, ˃540 U/L peroxidase, ˃400 U/L glycerol kinase, stabilizers, and preservatives.
- Standard: A solution containing 200 mg/dL glycerol and preservative.

Step-by-step method details (experimental procedure)
Procedure
The above contents are pipetted into clean dry test tubes labeled blank (B), standard (S), and test (T).

	(B)	(S)	(T)
Working reagent	1 mL	1 mL	1 mL
Serum sample	—	—	10 ul
Standard	—	10 ul	—

All the above constituents are mixed and incubated at 37°C for 10 min. The O.D of test (T) and standard (S) against blank (B) on a spectrophotometer at 520 nm is measured.

$$\text{Triglycerides concentration mg \%} = \frac{A \text{ of}(T)}{A \text{ of}(S)} \times 200$$

where A = Absorbance, S = Standard, T = Test.
The triglyceride was expressed as mg/dL of blood.

Expected outcomes
44−148 mg/dL (0.50−1.67 mmol/L)9. Due to a wide range of conditions (dietary, geographical, age, etc.) believed to affect normal ranges, it is recommended that each laboratory establish its own reference range.

Advantages
The enzymatic assay of cholesterol proved to be sensitive and precise. In comparison to other methods of cholesterol determination, it has the advantage of being rapid and simple. Most clinical and most research measurements of cholesterol use the enzymatic method. The enzymatic assay is specific, does not require corrosive chemicals, and is easily adapted for automation. The assay is generally performed as an end point or a kinetic method. Advantages of the kinetic method over the end point method are shorter analysis time, reduced effects of interfering substances, and elimination of sample blank measurement.

Limitations
The color reactions of cholesterol produce a number of products with varying absorptivities. This is an undesirable situation in quantitative analysis, and indicates the need to use these reactions under strictly controlled conditions. Furthermore,

these reactions are very nonspecific and require some degree of purification of the analyte before they can be applied successfully to the quantitative measurement of cholesterol in biological tissues or fluids.

Safety considerations and standards

Warm up working solution to the corresponding temperature before use. The reagent and sample volumes may be altered proportionally to accommodate different spectrophotometer requirements. Valid results depend on an accurately calibrated instrument, timing, and temperature control.

 • The reagent blank will not exceed an absorbance of 0.06 but don't use the reagent if it is turbid or if the absorbance is greater than 0.2 at 500 nm.

References

Demacker, P.N., Hijmans, A.G., Vos-Janssen, H.E., Van't Laar, A., Jansen, A.P., 1980. A study of the use of polyethylene glycol in estimating cholesterol in high-density lipoprotein. Clinical Chemistry 26 (13), 1775–1779.

Flegg, H.M., 1973. Ames award lecture 1972. An investigation of the determination of serum cholesterol by an enzymatic method. Annals of Clinical Biochemistry 10 (1–6), 79–84.

Fossati, P., Prencipe, L., 1982. Serum triglycerides determined colorimetrically with an enzyme that produces hydrogen peroxide. Clinical Chemistry 28 (10), 2077–2080.

Estimation of creatinine

Chapter outline

Principle:

The assay of creatinine is based on the reaction of creatinine with alkaline.

Creatinine + Picric Acid Alkaline → Creatinine − Picrate complex (yellow orange)

Materials and equipment (materials and reagents)

16×100 mm test tubes
Creatinine standard
Picric acid reagent
Controls
Sodium tungstate
Patient samples
Sulfuric acid
Spectrophotometer
Distilled water
Sodium hydroxide
Table top centrifuge

Basic Life Science Methods. https://doi.org/10.1016/B978-0-443-19174-9.00008-8

Creatinine buffer: A solution of sodium hydroxide containing at least 95 mmol/L sodium hydroxide.

Picric acid reagent: A solution of picric acid containing at least 25.8 mmol/L picric acid.

Creatinine standard: A solution containing 2.0 mg/dL creatinine and a preservative.

Reagent preparation:

Buffer solution is brought to room temperature. One vial of acid reagent is reconstituted with equal volume of buffer solution (working solution).

Step-by-step method details (experimental procedure)
Procedure

All the above reagents are pipetted into clean dry test tubes labeled standard (S) and test (T).

	(S)	(T)
Working solution Prewarm at 25, 30, or 37°C For 2 min and add	1.0 mL	1.0 mL
Standard	100 ul	—
Sample	—	100 ul

The above constituents are mixed well. The absorbance of test (T) and standard (S) on a spectrophotometer at 505 nm is measured.

$$\text{Creatinine in sample (mg / dL)} = \frac{\text{A of (T)}}{\text{A of (S)}} \times \text{Conc. of standard 1.5 mg/dL}$$

where A = absorbance, S = standard, T = test.

The creatinine is expressed by mg/dL.

Expected outcomes

A two-point calibration for serum creatinine combining only one serum standard with creatinine solution matching standard' sallow matrix-present interferents present results well comparable with enzymatic determination, providing the standard is attested/linked to the reference measurement procedure.

Normal Value: Serum: Male, 0.9–1.4 mg/dL, Female, 0.8–1.2 mg/dL.

Urine: Male, 0.4–1.8 g/24 h, Female, 0.35–1.6 g/24 h.

Each laboratory should establish its own normal range representing its patient population.

Advantages

Relatively cheap.

Limitations

a. Affected significantly by interfering molecules.
b. Poor specificity and sensitivity.
c. The nonspecific interaction between the small molecular weight substances present in the serum samples with alkaline picrate leads to the over estimation of serum creatinine.

Safety considerations and standards

1. Do not freeze the reagents.
2. During assay specified temperature has to be maintained.
3. Do not pipette the reagent by mouth.
4. Use clean glassware free from dust or debris.

Alternative methods/procedures

Other techniques for estimation of creatinine are ID-GC/MS method, enzymatic creatinine assay, etc.

Estimation of blood urea

Chapter outline

Principle:

Urea in the sample consumes by means of the coupled reactions described below, NADH that can be measured by spectrophotometry.

$$\text{Urea} + H_2O \xrightarrow{\text{Urease}} 2NH_4^+ + CO_2$$

$$NH_4 + NADH + H \quad 2- \text{oxoglutarate} \xrightarrow{\text{Glutamate Dehydrogenase}} \text{Glutamate} + NAD^+$$

Materials and equipment

Spectrophotometer
Pipette
Stop watch
Sample

Basic Life Science Methods. https://doi.org/10.1016/B978-0-443-19174-9.00009-X

Reagent: A solution containing a buffer (pH 8.0 at 25°C), 14 mmol/L 2-oxoglutarate, >1 KU/L GLDH, >5KU/L urease, stabilizers, and a preservative. Reagent NADH: A solution containing 0.18 mmol/L NADH and a preservative.

Reagent Preparation:

Working reagent—The contents of reagent B vial is transferred into a reagent A vial bottle. Solution is mixed gently, other volumes are prepared in the proportion 4 mL reagent A + 1 mL reagent B.

Step-by-step method details
Procedure

1. The working reagent is brought to room temperature before beginning test.
2. The following solutions are pipetted into a cuvette.

Working reagent	1.5 mL
Standard (S) or sample	10 ul

3. Mix the contents and start—stop watch,
4. The absorbance is recorded at 340 nm after 30 s (A 1) and after 90 s (A 2) by the formula mentioned below:

$$\frac{A1 - A2_{sample}}{A1 - A2_{standard}} \times C_{standard} \times \text{Sample dilution factor} = \text{urea mg/dL}$$

The urea is expressed by mg/dL.

Expected outcomes

The urea molecule contains two nitrogen atoms. So, the concentration of urea is expressed as blood urea nitrogen (BUN). The conversion of blood urea nitrogen (BUN) value to the blood urea is done by the following formula:

Value of blood urea = BUN × 2.14. Hence, the normal BUN = 6—22 mg/100 mL of serum.

Serum/plasma urea 2.5—7.8 mmol/L.

Advantages

The main advantage is rapid and suitable for routine analytical purposes. Precision and accuracy are good; sensitivity is high for an activated acid reagent up to about 1 week old, and thereafter decreases.

Limitations

It does not discriminate between the sample of interest and contaminants that absorb at the same wavelength. It is hardly simple enough for routine clinical pathology laboratories.

Safety considerations and standards

Observe universal precautions; wear protective gloves, laboratory coats. Place disposable plastic, glass, and paper (pipette tips, gloves, etc.) that contact plasma and any residual sample material in a biohazard bag and keep these bags in appropriate containers until disposal by maceration chlorination. Wipe down all work surfaces with Germicidal Disposable Wipe when work is finished. Handle acids and bases with extreme care; they are caustic and toxic. Handle organic solvents only in a well-ventilated area or, as required, under a chemical fume hood. Separated serum or plasma should not remain at $+15$ to $+30°C$ longer than 8 h. If assays are not completed within 8 h, serum or plasma should be stored at $+2$ to $+8°C$. If assays are not completed within 48 h, or the separated sample is to be stored beyond 48 h, samples should be frozen at -15 to $-20°C$. Frozen samples should be thawed only once. Analyte deterioration may occur in samples that are repeatedly frozen and thawed.

Alternative methods/procedures

Other techniques for estimation of blood urea are Jung urea assay, NMR, Berthelot's method, Dam Method, Fearon method, etc.

Estimation of blood glucose by GOD—POD method

Chapter outline

Principle: B-D glucose is oxidized to produce B-D gluconic acid and H_2O_2. The H_2O_2 is oxidatively coupled with G-amino antipurine and phenol substrate. Phenol substrate in the presence of peroxidase yields quinonimine dye. The amount of colored complex is propositional to glucose concentration and it is photometrically measured at 570 nm (Trinder, 1969).

$$\text{Glucose} + O_2 + H_2O \xrightarrow{\text{GOD}} \text{Gluconic acid} + H_2O_2$$

$$H_2O_2 + \text{Phenol} + \text{4-aminoantipyrine} \xrightarrow{\text{POD}} \text{Red Quinoneimine complex} + H_2O$$

Basic Life Science Methods. https://doi.org/10.1016/B978-0-443-19174-9.00010-6

Materials and equipment (materials and reagents)

Spectrophotometer (505 nm),
Vortex mixer
Micropipettor
Incubator
Glucose reagent: A solution containing a buffer (pH 7.5 at 25°C), 0.25 mmol/L 4-aminoantipyrine, 10 mmol/L phenol, '20,000 U/L glucose oxidase, '2000 U/L peroxidase and preservatives.
Glucose standard: Solution containing 100 mg/dL glucose and preservatives.

Step-by-step method details (experimental procedure)
Procedure

The test tubes labeled Blank (B), Standard (S), and Test (T) were pipetted as follows:

	(B)	(S)	(T)
Glucose reagent	1 mL	1 mL	1 mL
Glucose standard	–	0.01 mL	–
Serum sample	–	–	0.01 mL

The solutions are mixed and incubated for 10 min at 37°C. The absorbance of Test (T), standard (S), and Blank (B) at 505 nm was calculated by the following formula:

$$\text{Glucose Conc. in mg / dL} = \frac{\text{Absorbance of T}}{\text{Absorbance of S}} \times 100$$

The glucose levels were expressed by mg/dL.

Expected outcomes

Plasma glucose concentration in given unknown blood sample = —————— mg/dL.

Normal range blood

Random Blood Sugar: < 140 mg/dL.
Fasting Blood Sugar: 70–110 mg/dL.
Postprandial Blood Sugar: <140 mg/dL.

- CSF: 40–70 mg/dL (1/3 of plasma glucose)
- Urine: Absent

Advantages

The GOD-POD method is linear (up to 500 mg/dL), sensitive (detection limit 0.3 mg/dL), simple (requires 10 µL of sample to be incubated for 30 min with single reagent at room temperature) and requires simple instrumentation (the absorbance to be read between 505 and 550 nm).

Limitations

Very high glucose values with glucometer do not accurately reflect actual plasma glucose levels; but it overestimates glucose results. So, the routine practice of performing only single testing with glucometers can lead to misdiagnosis. So, readings obtained using glucometers especially at the critical hyperglycemic levels should be cautiously interpreted and verified with centralized laboratory.

Safety considerations and standards

Separate serum or plasma from red blood cell immediately after blood collection to avoid the glycolysis. Serum and plasma samples must be clarified. There may be a certain stimulative effect or toxicity while the reagents containing preservatives and stabilizer, please do not directly contact the skin and eyes. Rinsing with abundant water once the contact and do not swallow.

Alternative methods/procedures

Other techniques for Modified Folin Wu method • O-Toluidine method • Hexokinase method glucose oxidase assay, Nelson—Somogyi method, etc.

Reference

Trinder, P., 1969. Estimation of blood glucose by GOD-POD method. Annals of Clinical Biochemistry 6, 24.

Determination of albumin and globulin in plasma

11

Chapter outline

Principle: Albumin consists of approximately 60% of the total proteins in the body, the outer major part being globulin. It is synthesized in the liver and performs many like:

➢ Acts as a buffer in blood.
➢ Maintains osmotic pressure in blood.
➢ Acts as a carrier of hydrophobic substances like steroid hormones, bilirubin, fatty acids, etc.
➢ Acts as a source of amino acids for the synthesis of new proteins by cells.

Albumin binds with the dye bromocresol green in a buffered medium to form a green-colored complex. The intensity of the color formed is directly proportional to the amount of albumin present in the sample.

Materials and equipment (materials and reagents)

BCG reagent, albumin standard, bovine serum albumin (4 g/dL), Eppendorf tubes, and stand.

Step-by-step method details

Take five Eppendorf tubes and label them as blank (B), Standard (s) standard′ (s′), test (T), and test′ (T′).
    Pipette into the test tube as shown in table.
    Mix well and incubate at room temperature for 5 min.
    Measure absorbance of (S): (S′): (T): (T′): against blank (B).
    Note down the readings.

Tube	Reagent (mL)	Bsa standard	Sample	dH$_2$0	O.D	Average
Blank	1	–	–	10 µL	0	0
Standard (s)	1	10 µL	–	–	0.539	$\frac{0.539+0.506}{2}$
Standard' (s')	1	10 µL	–	–	0.506	$= 0.5225$
Test (T)	1	–	10 µL	–	0.518	$\frac{0.518+0.5557}{2}$
Test' (T')	1	–	10 µL	–	0.557	$= 0.5375$

$$\text{Albumin(g/dL)} = \frac{\text{Absorbance of Test}}{\text{Absorbance of Standard}} \times 4$$

$$\text{Albumin(g/dL)} = \frac{0.5375}{0.5225} \times 4 = \textbf{4.11g/dL}$$

$$\text{Globulin (g/dL)} = \text{Total protein} - \text{Albumin}$$

$$= 8.5 - 4.11 = 4.39 \text{ g/dL}$$

$$\text{A/G ratio} = \text{Albumin (g/dL)/Globulin (g/dL)}$$

$$\text{A/G ratio} = 4.11/4.39 = 0.936$$

Calculations

$$\text{Albumin(g/dL)} - \frac{\text{Absorbance of Test}}{\text{Absorbance of Standard}} \times 4$$

$$\text{Globulin (g/dL)} = \text{Total protein (g/dL)} - \text{Albumin (g/dL)}$$

$$\text{A/G ratio} = \frac{\text{Albumin(g/dL)}}{\text{Globulin(g/dL)}}$$

Quantitative determination of homocysteine

12

Chapter outline

Principle:

The quantitative estimation of Homocysteine was done by using enzyme immunoassay in human serum using a standard diagnostic test kit. The Axis Hcy Enzyme Immunoassay (EIA) kit (UK) is based on ELISA method.

Homocysteine Enzyme Immunoassay (EIA) is an enzyme immunoassay for the determination of Hcy in blood. Protein-bound Hcy is reduced to free Hcy and enzymatically converted to S-Adenosyl-Homocysteine (SAH) in a separate procedure prior to the immunoassay. The enzyme is specific for the L-form of homocysteine, which is the only form present in the blood.

Reduction:

Hcy, mixed disulfide, and protein bound forms of Hcy in the sample are reduced to free Hcy by use of Dithiothreitol (DTT).

Prot-SS-Hcy, R1-SS-Hcy, and Hcy-SS-Hcy are reduced to Homocysteine.

Materials and equipment (materials and reagents)

- Precision micropipettes to dispense 25, 100, 200, and 250 μL.
- Disposable pipette tips.
- Volumetric flask—50 and 600 mL.
- Washer and reader for microtitre plates.

- ELISA reader.
- Microwell plate reader with a filter set at 450 nm and an upper OD limit of 450 nm.
- Reagent-A: Assay buffer—Phosphate buffer and sodium azide—54 mL.
- Reagent-B: Adenosine DTT—Adenosine dithiothreitol, citric acid—3.5 mL.
- Reagent-C: SAH-hydrolase-recombinant S-adenosyl-L-homocysteine hydrolase, Tris buffer, glycerol, methylparaben—3.5 mL.
- Reagent-D: Enzyme inhibitor—Merthiolate, phosphate buffer—55 mL.
- Reagent-E: Adenosine deaminase—Adenosine deaminase, phosphate buffer, sodium azide, BSA, phenol-red-dye—55 mL.
- Reagent-F: a-SAH-antibody—Monoclonal mouse—anti-adenosyl-L-homocysteine antibody, BSA, merthiolate—25 mL.
- Reagent-G: Enzyme conjugate—Rabbit antimouse—antibody enzyme conjugate, BSA, Horse radish peroxidase, blue dye—15 mL.
- Reagent-H: Substrate solution—N-methyl-2-pyrrolidin, propyleneglycol—15 mL.
- Reagent-S: Stop solution—0.8M sulfuric acid—20 mL.
- Buff Wash: Wash buffer—Phosphate buffer, merthiolate, Tween 20, BSA—60 mL.
- CAL-1 to CAL-6: Calibrators-S-adenosyl-L-homocysteine (2, 4, 8, 15, 30, 50 µmol/L) in buffer with preservative—1.5 mL per each calibrator.
- **Microtitre Strips**: Microtitre strips coated with S-adenosyl-L-homocysteine.

Step-by-step method details (experimental procedure)
Procedure

1. Sample pretreatment solution are made more than 1 h prior to the start of assay. Volume needed per 10 samples

   4.5 mL—Reagent-A
   0.25 mL—Reagent-B
   0.25 mL—Reagent-C
   All reagents are mixed well and kept at 37°C.
2. Calibrators and samples are diluted in a plastic or glass tubes.
   25 µL Calibrator/Sample/Control.
   +500 µL Sample pretreatment solution.
   Mix well and incubate for 30 min at 18−20°C. Proceed with Step 3 before samples have cooled.
3. Add 500 µL Reagent-D
   Mix well. Incubate for 5 min at 18−20°C.
4. 500 µL Reagent-E is added
   Mix well. Incubated at 18−20°C.

Microtiter procedure

25 μL diluted calibrator/sample/control is pipetted from step 4 into the SAH-coated microtiter strips. 200 μL of Reagent-F is added into each well and incubated for 30 min at 18−20°C. The enclosed lid is used during all incubations. Washing with diluted Wash Buffer (WASH BUFF + purified water), 4 times with 400 m. After washing, empty the wells on paper towels. Add 100 μL Reagent-G to each well. Incubate for 20 min at 18−20°C. Wash 4 times 350 μL of diluted Wash Buffer (WASH-BUFF + Purified water). After washing, empty the wells on paper towels. Add 100 μL Reagent-H to each well. Incubate for 10 min at 18−20°C. Add 100 μL Reagent-S to each well. Shake and read at 450 nm within 15 min.

Expected outcomes

The normal range of homocysteine is 5−15 μmol/L. When the level is between 16 and 30 μmol/L, it is classified as moderate, 31−100 μmol/L is considered intermediate and a value above 100 μmol/L is classified as severe.

Advantages

The method has a practical advantage over other methods since it does not involve cumbersome, expensive, and sophisticated equipment, thus making it more feasible for laboratories. The method avoids the use of radioisotopes and chromatographic separations and relies on enzymatic conversion of homocysteine to S-adenosyl-L-homocysteine, followed by quantification of S-adenosyl-L-homocysteine by an enzyme-linked immunoassay in microtiter format. It helps to find if you have deficiency in vitamin B12, B6, or folic acid. Help diagnose homocystinuria, a rare, inherited disorder that prevents the body from breaking down certain proteins.

Limitations

Labor-intensive and expensive to prepare antibody because it is a sophisticated technique, and expensive culture cell media are required to obtain a specific antibody. (ii) High possibility of false-positive or false-negative results because of insufficient blocking of the surface of microtiter plate immobilized with antigen. (iii) Antibody instability because an antibody is a protein that requires refrigerated transport and storage.

Safety considerations and standards

Observe universal precautions; wear protective gloves, laboratory coats, and safety glasses during all steps of this method. Discard any residual sample material by autoclaving after analysis is completed. Place disposable plastic, glass, and paper (pipet tips, autosampler vials, gloves, etc.) that contact plasma in a biohazard autoclave bag and keep these bags in appropriate containers until sealed and autoclaved. Wipe down all work surfaces with 10% bleach solution when work is finished. Handle acids and bases with extreme care; they are caustic and toxic. Handle organic solvents only in a well-ventilated area or, as required, under a chemical fume hood. For best results, a fasting sample should be obtained. Specimens for total homocysteine analysis may be fresh or frozen plasma.

Estimation of serum bilirubin

Chapter outline

Principle: In the determination of total bilirubin, bilirubin is coupled with diazotized sulfanilic acid in the presence of surfactant to produce azobilirubin which has maximum absorbance at 546 and 630 nm. Direct bilirubin in presence of diazotized sulfanilic acid forms a red-colored azo compound in acidic medium which has maximum absorbance at 546 and 630 nm (diazo method of Pearlman and Lee, 1974).

Reagents required

R$_1$:
1. Sulfanilic acid   5 mmol/L
2. Hydrochloric acid   100 mmol/L
3. Surfactant   QS

R2:
1. Sulfanilic acid   50 mmol/L
2. Hydrochloric acid   100 mmol/L

Basic Life Science Methods. https://doi.org/10.1016/B978-0-443-19174-9.00013-1
Copyright © 2023 Elsevier Inc. All rights reserved.

R3:
Sodium nitrite   144 mmol/L

Materials required

Analyzer/photometer, pipettes, distilled water, bilirubin (total and direct) test kit and other lab equipment.

Reagent preparation

The working reagent is prepared as per user's need.

Test	Volume of working reagent (mL)	R1	R2	R3 (mL)
Bilirubin total	5	5 mL	... ...	0.25
Bilirubin direct	5	... ...	5 mL	0.25

Assay procedure
Total bilirubin end point (bichromate)

Wavelength	Hg 545 nm and 630 nm
Temperature	37°C
Factor	23
Cuvette	1 cm light path
Node	End point bichromatic

Assay procedure
Direct bilirubin end point (bichromatic)

Wavelength	Hg 545 and 630 nm
Temperature	37°C
Factor	23
Cuvette	1 cm light path
Node	End point bichromatic

Calibration: For the calibration of automated photometric systems, use the commercially available calibrator (calibration factor of 23 may be used for total bilirubin and calibration factor of 17 for direct bilirubin).

Quality control: To ensure adequate quality, use of the commercially available control sera is recommended.

Measuring range: The test has been developed to determine bilirubin concentrations within a measuring range from 0.05 to 20 mg/dL (0.85–345 umol/L). When values exceed higher limit of the range, such samples should be diluted 1 + 1 with NaCl solution (9 g/L) and the result multiplied by 2.

Conversion factor: mg/dL × 17.1 = umol/L.

Sensitivity/limit of detection: The lower limit of detection is 0.05 mg/dL (0.85 umol/L).

Specimen: Serum free from hemolysis.

Reference range: Expected values.

	Total bilirubin	Direct bilirubin
Adult	1.1 mg/dL (18.81 umol/L)	0.25 mg/dL (4.27 umol/L)

PREMATURE	
0–1 day	<8.0 mg/dL (136.8 umol/L)
1–2 days	<12 mg/dL (205.2 umol/L)
3–5 days	<16 mg/dL (273.6 umol/L)
Above 5 days	0.3–1.2 mg/dL (5.13–20.52 umol/L)

NOTE: It is recommended that each laboratory should assign its own reference range.

Another procedure

Aromatic primary amine, 5 mmol/L; 500 μmol of aromatic primary amine was dissolved in 100 mL of 0.176 mol/L hydrochloric acid solution. Sodium nitrite solution, 72 mmol/L; 0.5 g of sodium nitrite was dissolved in 100 mL of distilled water, and stored in a refrigerator. This reagent was prepared every week.

Diazo reagent 0.6 mL of the sodium nitrite solution was added to 10 mL of the aromatic primary amine solution. This reagent was used within 1 h of its preparation. Bilirubin solution, 170 μmol/L; 10.0 mg of unconjugated bilirubin was dissolved in 100 mL of chloroform, and stored in a refrigerator.

Preparation of the water-soluble photoproduct 10 mL of the bilirubin solution was irradiated by a fluorescent lamp at an intensity of 1500 lux during 48 h. Then, 10 mL of distilled water was added to this solution, and tracted.

Diazo reaction of the water-soluble photoproduct 1.0 mL of the diazo reagent was added to 5.0 mL of the photoproduct extract, and immediately the absorbance at 540 nm was measured at 25°C against a reagent blank which consisted of 1.0 mL of the diazo reagent and 5.0 mL of distilled water, using a Shimadzu double beam UV 200 spectrophotometer. The rate constant of the reaction was calculated from these results.

Determination of serum bilirubin serum total bilirubin; In a test tube, containing 0.1 mL of serum and 3.9 mL of 50% methanol, 1.0 mL of the diazo reagent was added and the solution was mixed well. Then the reaction mixture was allowed to stand for 30 min at 25°C. The absorbance was then measured at 540 nm with a Hitachi 101 spectrophotometer against a reagent blank, consisting of 4.0 mL of 50% methanol and 1.0 mL of the diazo reagent. Serum direct bilirubin; In a test tube containing 0.1 mL of serum and 3.9 mL of distilled water, 1.0 mL of the diazo reagent was added and the solution was mixed well. Then the reaction mixture was allowed to stand for 10 min at 25°C. The absorbance was measured at 540 nm against a reagent blank, consisting of 4.0 mL of distilled water and 1.0 mL of the diazo reagent.

Calibration curve In six tubes, 0.1, 0.2, 0.3, 0.4, and 0.5 mL of the unconjugated bilirubin solution was diluted to 4.0 mL with methanol. To each tube, 1.0 mL of the diazo reagent was added and mixed well. Then each reaction mixture was allowed to stand for 10 min at 25°C. The absorbance was measured at 540 nm against a reagent blank, consisting of 4.0 mL of methanol and 1.0 mL of the diazo reagent.

Results

Absorbance change of unconjugated bilirubin and patients' serum on irradiation with a fluorescent lamp. The effect of irradiation with a fluorescent lamp at 1500 lux on the decomposition of unconjugated and patients' serum bilirubin will be examined.

Reference

Pearlman, F.C., Lee, R.T., 1974. Detection and measurement of total bilirubin in serum, with use of surfactants as solubilizing agents. Clinical Chemistry 20 (4), 447–453.

Estimation of diagnostic enzyme ALT

14

Chapter outline

Methodology: IFCC method without pyridoxal phosphate, kinetic, UV.
    Principle: Kinetic determination of the GPT activity:
    L-Alanine + alpha ketoglutarate $\rightarrow$ GPT   Pyruvate + L-Glutamate
    Pyruvate + NADH + H^+ $\rightarrow$ LDH   L-Lactate + NAD^+

Reagents

Components and concentrations

R1:

    Tris buffer   100 mmol/L

L-alanine   500 mmol/L
Lactate dehydrogenase   >1200 u/L

R2:

Alpha ketoglutarate   1500 mmol/L
NADH   0.18 mmol/L
Preservative and Stabilizer

Materials required

ALT/SGPT test kit, NaCl solution 9 g/L, general laboratory equipment, analyzer/photometer, pipettes, etc.

Regent preparation
One reagent procedure

- Mix four volumes of reagent R1 with one volume of reagent R2.
- Stability of working reagent solution: 4 weeks at 2 to 8°C.

   Two reagent procedures: Ready to use reagents.

Specimen

- Serum free from hemolysis, heparinized or EDTA plasma.
- Stability in serum/plasma.
- Sera are stable 24 h at 20 to 25°C, 28 days at 4°C.
- Discard contaminated specimen.

Assay procedure
Two reagent procedure

Wavelength   Hg 340 nm
Temperature   37°C
Optical path   1 cm
Mode   Kinetic

- Bring all the contents of the kit to room temperature prior to use.
- Read rate of change of absorbance of sample against distilled water or air.
- Label the test tube as sample, control and pipette with respective test tube the reagent, sample, control sample as per the table given below.

Sample/control

Reagent R1	800 uL
Reagent R2	200 uL

- Mix and incubate at 37°C for 2 min then add sample/control 100 uL.
- Mix and after a 60 s incubation at 37°C, measure the change of absorbance for minute (A/min) during 180 s.

Assay procedure 2

One reagent procedure

- Label the test tube as sample, control, and pipette into respective test tube the reagent, sample, control sample as per the table given below.

Sample/control

Working reagent	1000 uL
Sample/control	100 uL

Prewarm working reagent at 37°C for 2 min prior to addition of sample.

- Mix and after a 60 s incubation at 37°C, measure the change of absorbance for minute (A/min) during 180 s.

Calculations

At 340 nm with one reagent procedure and two reagent procedures for 1 cm path light cuvette.

Activity of sample (U/L): (A/min) $\times$ 1746.

Calibrations: For the calibration of automated photometric systems, use of the commercially available calibrator is recommended.

Quality control

To ensure adequate quality, use of the commercially available control sera is recommended.

Performance characteristics

- **Measuring range:**
  The test tube has been developed to determine GPT/ALT activity within a measuring range from 5 to 400 U/L. When values exceed higher limit of the range, such samples should be diluted 1 + 1 with NaCl solution (9 g/L) and the result multiplied by 2.
- **Sensitivity/limit of detection:**
  The lower limit of detection is 5 U/L.
- **Reference range:**
  Men   0—40 U/L   Women   0—32 U/L
  It is recommended that each laboratory should assign its own reference range.

Temperature conversion factors

To convert result to other temperatures, multiply by factor shown in table.

Assay temperature	Conversion factor to		
	25°C	30°C	37°C
25°C	1.00	1.37	2.08
30°C	0.73	1.00	1.54
37°C	0.48	0.65	1.00

Estimation of diagnostic enzyme AST

Chapter outline

Methodology: IFCC method without pyridoxal phosphate, kinetic, UV

Principle: Aspartate aminotransferase abbreviated as AST is an enzyme which is found mainly in the liver and red blood cells, heart cells, muscle tissue and other organs, like pancreas and kidneys. The levels of AST in serum diagnose body tissues like heart and liver are injured or not. During the reaction the AST catalyzes the reversible transamination of L-aspartate and α-ketoglutarate to oxaloacetate and L-glutamate. The oxaloacetate is then reduced to malate in the presence of malate dehydrogenase with the concurrent oxidation of NADH to NAD. The rate of change in absorbance at 340 nm over a fixed-time interval is monitored which is directly proportional to the AST activity in the sample.

$$\text{L – Aspartate + alpha ketoglutarate} \xrightarrow{\text{GOT}} \text{Oxaloacetate + L – Glutamate}$$

$$\text{Oxaloacetate + NADH + H}^+ \xrightarrow{\text{MDH}} \text{L- Malate + NAD}^+$$

Reagents
Components and concentrations

R1:

Tris buffer	80 mmol/L
L-aspartate	240 mmol/L

Lactate dehydrogenase >600 u/L
Malate dehydrogenase >600 u/L

R2:

Alpha ketoglutarate	12 mmol/L
NADH	0.18 mmol/L

Preservative and stabilizer

Materials required

AST/SGOT test kit, NaCl solution 9 g/L, general laboratory equipment, analyzer/photometer, pipettes, etc.

Regent preparation
One reagent procedure

- Mix four volumes of reagent R1 with one volume of reagent R2.
- Stability of working reagent solution: 4 weeks at 2−8°C.

  Two reagent procedures: Ready to use reagents

Specimen

- Serum free from hemolysis, heparinized or EDTA plasma.
- Stability in serum/plasma.

- Sera are stable 24 h at 20–25°C, 28 days at 4°C.
- Discard contaminated specimen.

Assay procedure

1. Two reagent procedure:

Wavelength	Hg 340 nm
Temperature	37°C
Optical path	1 cm
Mode	Kinetic

- Bring all the contents to room temperature prior to use.
- Read rate of change of absorbance of sample against distilled water or air.
- Label the test tube as sample, control and pipette with respective test tube the reagent, sample, control sample as per the table given below

Sample/control

Reagent R1	800 uL
Reagent R2	200 uL.

- Mix and incubate at 37°C for 2 min then add sample/control 100 uL.
- Mix and after a 60 s incubation at 37°C, measure the change of absorbance for minute (A/min) during 180 s.

Assay procedure 2
One reagent procedure

- Label the test tube as sample, control and pipette into respective test tube the reagent, sample, control sample as per the table given below

Sample/control

Working reagent	1000 uL
Sample/Control	100 uL

Prewarm working reagent at 37°C for 2 min prior to addition of sample.

- Mix and after a 60 s incubation at 37°C, measure the change of absorbance for minute (A/min) during 180 s.

Calculations

At 340 nm with one reagent procedure and two reagent procedures for 1 cm path light cuvette.

Activity of sample (U/L): (A/min) X 1746

Calibrations: For the calibration of automated photometric systems, use of the commercially available calibrator is recommended.

Quality control

To ensure adequate quality, use of the commercially available control sera is recommended.

Performance characteristics

- **Measuring range:**

  The test tube has been developed to determine GOT/AST activity within a measuring range from 5 to 400 U/L. When values exceed higher limit of the range such samples should be diluted $1 + 1$ with NaCl solution (9 g/L) and the result multiplied by 2.

- **Sensitivity/limit of detection:**

  The lower limit of detection is 5 U/L.

- **Reference range:**
  Men: 0–38 U/L   Women 0–31 U/L

  It is recommended that each laboratory should assign its own reference range.

- **Temperature conversion factors:**

  To convert result to other temperatures, multiply by factor shown in table

Assay temperature	Conversion factor to		
	25°C	30°C	37°C
25°C	1.00	1.37	2.08
30°C	0.73	1.00	1.54
37°C	0.48	0.65	1.00

Protein estimation by Lowry's method

16

Chapter outline

Principle: In the presence of alkaline copper reagents aromatic amino acids like tyrosine, tryptophan present in protein sample mainly react with phosphotungstic and phosphomolybdic acid and give blue color at λ max at 660 nm.

Materials and equipment (materials and reagents)

- Standard—BSA (bovine serum albumin)
- Reagent A—2% Na_2Co_3 + 0.1N NaOH [50 mL]
- Reagent B—2% $CuSO_4.5H_2O$ [5 mL]
- Reagent C—2% potassium sodium tartarate [5 mL]
- Reagent D—49.5 mL reagent A + 0.25 mL reagent B + 0.25 mL Reagent C
- Water bath
- Spectrophotometer
- Test tubes
- Test tube stand
- Pipettes
- Tips
- Tip box
- Beaker.

Step-by-step method details (experimental procedure)

Procedure

1. Pipette out 0.1, 0.2, 0.3, 0.4, and 0.5 mL of the working standard into a series of test tube.
2. Pipette out 0.2 and 0.2 mL of the sample extract in two other test tubes.
3. Make up the volume to 1 mL with distilled water in all the test tubes. A tube with 1 ml of water serves as the blank.
4. Add 5 mL of reagent D to all test tubes, mix well, and incubate at room temperature in the dark for 30 min. Blue color will develop.
5. Add 0.5 mL FCR in dark place to all test tubes.
6. Incubate again for 30 min in dark at room temperature.
7. Take the reading at 630 nm.
8. Draw a standard graph and calculate the amount of protein in the sample.

Protein Estimation by Lowry's Assay.

Test tubs	Vol. of Std. (BSA) Or sample (μL)	Vol. of distilled water (μL)	Vol. of reagents D	Incubation time	Vol. of FC reagent	Incubation time	OD @ 630 nm
Blank	0	1000	5 mL		0.5 mL		00.00
Standard 1	100	900	5 mL		0.5 mL		0.42
Standard 2	200	800	5 mL		0.5 mL		0.53
Standard 3	300	700	5 mL		0.5 mL		0.65
Standard 4	400	600	5 mL		0.5 mL		0.66
Standard 5	500	500	5 mL		0.5 mL		0.67
Sample 1	200	800	5 mL		0.5 mL		1.30
Sample2	200	800	5 mL		0.5 mL		0.74
Sample3	200	800	5 mL		0.5 mL		1.22
Sample4	200	800	5 mL		0.5 mL		0.60

Expected outcomes

The concentration of protein can be calculated by plotting the graph of standard protein. The concentration of sample 1 (40% acetone precipitated) was 14.3 ug/mL and sample 2 (60% acetone precipitated) was 5.44 ug/mL, sample 3 (80% acetone precipitation) was 13.06 ug/mL and sample 4 (100% acetone precipitation) was 3.22 ug/mL.

Advantages

Lowry's assay has major advantages: (i) sensitive assay which requires no digestion of proteins; (ii) more specific and less interrupted by turbidity; (iii) convenience through stability of the reagent formulations; (iv) measurement of protein in both colorless and colored biological samples without compromising the sensitivity; and (v) assaying proteins at very low concentrations.

Limitations

The amount of color developed differs from protein to protein. It is less constant than the biuret reaction, but more constant than the absorbance at 280 nm. The color is not exactingly proportional to concentration.

Safety considerations and standards

Agents that acidify the solution such as strong acid or high concentrations of ammonium sulfate interfere with the assay as does EDTA which chelates copper or dithiothreitol or mercaptoethanol that cause the reduction of Cu^{2+}. Thus, it is critical that you remove these complicating reagents before carrying out an assay.

Alternative methods/procedures

Other techniques are colorimetric assays, bicinchoninic acid (BCA) assay, Bradford method, amino acid analysis, fluorescence assay.

Estimation of reducing sugar by using dinitro salicylic acid method

17

Chapter outline

Principle: Reducing sugars have the property to reduce many of the reagents. One such reagent is 3,5- dinitrosalicylic acid (DNS). 3, 5-DNS in alkaline solution is reduced to 3 amino 5 nitro salicylic acid. This method tests for the presence of free carbonyl group ($C{=}O$), the so-called reducing sugars. This involves the oxidation of the aldehyde functional group present in, for example, glucose and the ketone functional group in fructose. Simultaneously, 3,5-dinitrosalicylic acid (DNS) is reduced to 3-amino-5-nitrosalicylic acid under alkaline conditions, as illustrated in the equation below in Fig. 17.1:

Materials and equipment (materials and reagents)

- **Standard glucose solution:**

  0.01 g anhydrous glucose is dissolved in distilled water and then raised the volume to 10 mL with distilled water.
- **Dinitrosalicylic acid reagent:**

  Solution "A" is prepared by dissolving 6g of sodium potassium tartarate in about 10ml distilled water.

  Solution "B" is prepared by dissolving 0.2 g of 3, 5-dinitrosalicylic acid in 4 mL of 2N NaOH solution.

Basic Life Science Methods. https://doi.org/10.1016/B978-0-443-19174-9.00017-9

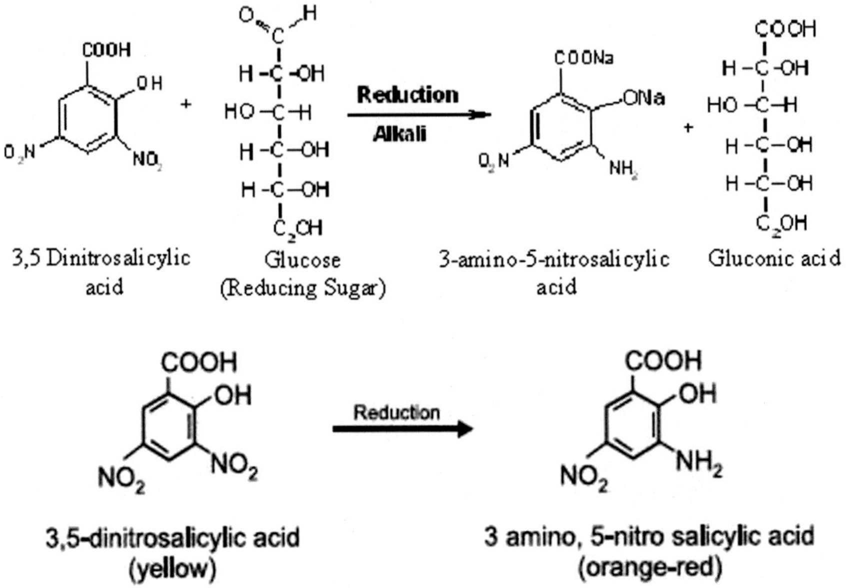

FIGURE 17.1

Color change during the reaction.

The dinitrosalycilate reagent is prepared by mixing solutions A and B and raising the final volume to 20 L with distilled water.

Step-by-step method details (experimental procedure)
Procedure

1. Pipette in the following reagents into a series of dry-clean and labeled test tubes and as indicated in the following table, take Section A.

	Section A			Section B
Tube no.	Stand. Glucose (mL)	H_2O (mL)	Dinitrosalicylic reagent (mL)	H_2O (mL)
1	0.0	1.0	2.0	7.0
2	0.2	0.8	2.0	7.0
3	0.4	0.6	2.0	7.0
4	0.6	0.4	2.0	7.0
5	0.8	0.2	2.0	7.0
6	1.0	0.0	2.0	7.0

2. After replacing the above-mentioned solutions as in section A in the labeled tubes, shake well and then place them in a boiling water bath for 5 min.

3. Cool the tubes thoroughly and then add 7.0 mL of distilled water to each tube as indicated in Section B of the previous table. Read the extinction (optical density) of the colored solutions at 540 nm using the solution in tube 1 as a blank (control).

(Note: All the tubes must be cooled to room temperature before reading since the extinction is sensitive to temperature change.)

4. Record the readings in Section B, and plot the relationship between the optical density and the concentration of glucose solution. See whether there is a linear relationship between the concentrations of glucose solutions and their corresponding optical densities (Fig. 17.2).

5. Prepare standard curves of the sugars provided and use them to estimate the concentration of the unknowns provided (Fig. 17.3).

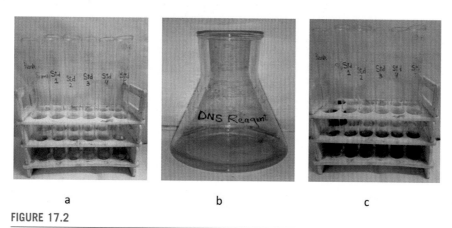

a   b   c

FIGURE 17.2

Sugar estimation by DNS method: (A) Before incubation, (B) Solution mixture, (C). After incubation.

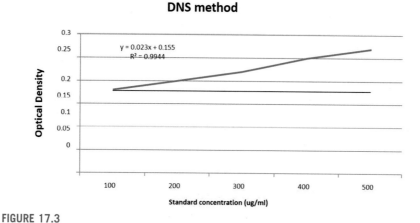

FIGURE 17.3

The concentration of sugar calculated by plotting the graph of standard protein.

Expected outcomes

Sugar estimation by DNS method

Test tubes	Vol. of std. or sample (mL)	Vol. of distilled water (mL)	Vol. of DNS reagent (mL)	Incubation time	OD @ 540 nm
Blank	0	1000	2 mL		00.00
Standard 1	100	900	2 mL		0.18
Standard 2	200	800	2 mL		0.20
Standard 3	300	700	2 mL		0.22
Standard 4	400	600	2 mL		0.25
Standard 5	500	500	2 mL		0.27
Sample 1	200	800	2 mL		1.39

The concentration of sugar was calculated by plotting the graph of standard protein. The sugar sample was isolated from soft drinks.

Let the concentration of reducing sugar in sample be x $Y = 0.023x + 0.155$

$1.39 = 0.023x + 0.155$

$X = 53.69 ug/mL$

Advantages

It is seen that DNSA methods have advantage of its simplicity, low cost, sensitivity, and adoptable during handling the analysis of a large number of samples at a time. The Dinitro Salicylic Acid (DNSA) assay is a simple and inexpensive method. If glucose is the only product, the DNSA assay can be used as an accurate analytical method for evaluating reducing sugars in both pure solutions and supernatants from enzymatic saccharifications of purified cellulosic.

Limitations

Because the DNSA assay has a low specificity, it is critical to run blanks diligently if the calorimetric results are to be interpreted correctly and accurately. Because DNSA reagent is destructive to polysaccharides, it cannot be used to measure pectinase activity against pectins. Overestimation is also associated with the DNS assay.

Safety considerations and standards

The test is sensitive to the room temperature, so, cool down all the samples to the room temperature before reading. Always keep the enzyme extract at 4°C. The dilution of the enzyme can be adjusted according to the color produced, but always remember the dilution factor while calculating the activity.

Alternative methods/procedures

Other techniques are Nelson—Somogyi Method/Folin Wu Method/Glucose Oxidase Method/Benedict's Method/Fehling's Method.

To perform quantitative determination of total protein in serum/plasma

Chapter outline

Principle: In an alkaline medium, total protein reacts with the copper of biuret reagent causing an increase in absorbance at 546 nm (530—570 nm or green filters) due to concentration of protein present in the sample.

$$\text{Total protein} + \text{Cu} \longrightarrow \text{Voilet complex}$$

Materials and equipment (materials and reagents)

- Reagent 1: Biuret reagent
- Reagent 2: Protein standard 6 g/dL
- Clean and dry glasswares
- Micropipettes
- Microtips
- Spectrophotometer
- Sample: Serum/Heparinized plasma (or) EDTA plasma

Basic Life Science Methods. https://doi.org/10.1016/B978-0-443-19174-9.00018-0

Step-by-step method details (experimental procedure)
Procedure

1. Mark the test tubes and pipette assays as mentioned below:

S. No.	Reagent	Blank	Standard	Test
1	Working reagent	1 mL	1 mL	1 mL
2	Standard	–	10 ul	–
3	Sample	–	–	10 ul

2. Mix well; incubate for 5 min at room temperature.

3. Measure the absorbance of standard and sample against blank reagent at 546 nm (530–570) nm.

4. The final color remains stable for 1 h.

Calculation:

Total protein concentration (gm/dL) = [O.D. sample/O.D. standard] * concentration of standard solution.

Expected outcomes

Quantitative determination of total protein in serum/plasma

S. No.	Cell name	Wavelength	Absorbance
1.	Blank	546 nm	0.00
2.	Standard	546 nm	0.03
3.	Sample	546 nm	0.03

Total protein concentration (gm/dL) = [O.D. sample/O.D. standard] * concentration of standard solution.

Protein concentration = [0.03/0.03]*5

= **5 gm/dL.**

Blank standard and sample were prepared according to the above table and their O.D where taken at 546 nm and the total protein concentration in the serum sample was calculated.

Advantages

The biuret assay is still frequently used because of simple analytical procedure, easy preparation of reagents, and when compared with other copper-based assays, this method is less susceptible to chemical interference.

Limitations

This method is not sensitive enough to measure lower protein concentrations found, for example, in cerebrospinal fluid.

Safety considerations and standards

Do not in any case freeze or expose reagent to high temperature as it may affect the performance. Before the assay bring all the reagents to room temperature. Avoid contamination of the reagents during the assay process. Use clean glassware free from dust or debris.

Alternative methods/procedures

Bradford (Coomassie Brilliant Blue), Lowry (Folin-Ciocalteau), Biüret, Pesce and Strande (Ponceau-S/TCA), and modified method of Schaffner-Weismann (Amido Black 10B) are some alternative methods.

CHAPTER

Enzyme activity using starch assay

19

Chapter outline

Enzymes are proteins that act as catalysts for biological reactions. Enzymes, like all catalysts, speed up the rate of reaction without being used up themselves. The amylase enzyme is responsible for hydrolyzing starch. In the presence of amylase, a sample of starch will be hydrolyzed to shorter polysaccharides, dextrins, maltose, and glucose. The hydrolysis of starch to glucose is catalyzed by amyloglucosidase. Glucose is phosphorylated by adenosine triphosphate (ATP) in the reaction catalyzed by hexokinase. Glucose-6-phosphate (G6P) is then oxidized to 6-phosphogluconate in the presence of nicotinamide adenine dinucleotide (NAD) in a reaction catalyzed by glucose-6-phosphate dehydrogenase (G6PDH).

Requirements

- ➢ Enzyme sample
- ➢ 1% starch solution
- ➢ Glucose
- ➢ Test tubes
- ➢ Pipette
- ➢ Pipette tips
- ➢ Water bath
- ➢ Centrifuge

Basic Life Science Methods. https://doi.org/10.1016/B978-0-443-19174-9.00019-2

Preparation

✔ 1% Starch Substrate Solution

Starch 0.2 gm

Sodium phosphate buffer (0.25M) 05 mL

} mix by stirring & gentle heating

✔ Glucose (1 mg/mL)10 mL
✔ DNS reagent 20 mL

Procedure

1. 1% of starch substrate is brought to reaction temperature (30°C).
2. 0.1 mL of starch substrate solution is added to each of test tube (without blank and sample ones).
3. Add 0.1 mL glucose solution to test tubes containing 0.2 mL of sample solution.
4. Add distilled water in each test tube to make volume 1 mL.
5. Add 1 mL DNS reagent in each test tube with gentle mixing.
6. Incubate for 10 min at 37°C in water bath and wait if black-reddish color will appear.

Observation and result

Yellow color solution was changed to red color after incubation that shows the presence of reducing sugar (Fig. 19.1).

A

B

FIGURE 19.1

Determination of enzyme activity: (A) Before incubation, (B) After incubation.

Determination of enzyme activity

The determination of enzyme activity was calculated by plotting the graph between standard concentration and optical density. The concentration of sample 1 was 4.50 ug/dL and sample 2 was 4.72 ug/dL (Fig. 19.2 and Table 19.1).

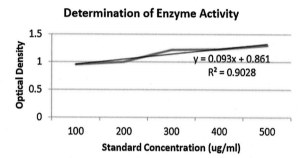

FIGURE 19.2

Graph represents the standard for Determination of Enzyme Activity.

Table 19.1 Determination of enzyme activity.

Test tube No.	Vol. of sample/ Std.	Distilled water (μL)	Vol. of DNS reagent	Incubation time	Distilled water	O.D at 540 nm
Blank	0 μL	1000	1 mL		7 mL	0.00
Standard 1	100 μL	900	1 mL		7 mL	0.95
Standard 2	200 μL	800	1 mL		7 mL	1.00
Standard 3	300 μL	700	1 mL		7 mL	1.22
Standard 4	400 μL	600	1 mL		7 mL	1.23
Standard 5	500 μL	500	1 mL		7 mL	1.30
Sample 1	200 μL Glucose sol^n +200 μL Sample	600	1 mL		7 mL	1.28
Sample 2	200 μL Glucose sol^n +200 μL Sample	600	1 mL		7 mL	1.30

To estimate protein by Bradford assay

20

Chapter outline

Principle: Bradford assay is a quantitative method for the estimation of enzymes. Coomassie Brilliant Blue Dye (CBB) in Bradford reagents with proteins sample give a blue color solution, where absorption shifts from 465 to 595 nm and we can take readings at 595 nm.

Materials and equipment (materials and reagents)

- Bradford reagents
- Crude sample
- Distilled water
- Autoclave
- Water bath
- Spectrophotometer
- Coomassie blue
- Ethyl alcohol
- Hydrogen phosphate

Basic Life Science Methods. https://doi.org/10.1016/B978-0-443-19174-9.00020-9

Bradford reagent preparation

- Take a pinch of CBB dye and dissolve it in 2 mL of 95% C_2H_5OH.
- Add 2 mL of 85% H_3PO_4 and make up the final volume 20 mL with distilled water.

Step-by-step method details (experimental procedure)
Procedure

1. Take 11 test tubes. In a test tube, take aliquots of standard BSA 100–500 μL.
2. Add distilled water in decreasing order in each test tube from 500 to 100 μL.
3. 2.5 mL of Bradford reagent is added in each tube.
4. The test tubes are incubated in dark at room temperature for 10 min.
5. Absorbance is calculated at 630 nm.

Expected outcomes

Bradford assay

Test tubes	Vol. of std. or sample (μL)	Vol. of distilled water (μL)	Vol. of Bradford reagent	Incubation time	OD at 630 nm
Blank	0	1000	2.5 mL	30 min	0.00
Standard 1	100	900	2.5 mL		0.06
Standard 2	200	800	2.5 mL		0.07
Standard 3	300	700	2.5 mL		0.13
Standard 4	400	600	2.5 mL		0.17
Standard 5	500	500	2.5 mL		0.21
Sample 1	200	800	2.5 mL		1.37
Sample 2	200	800	2.5 mL		1.56

Result

The concentration of protein is calculated by plotting the graph of standard protein (Fig. 20.1). The protein sample is isolated from apple juice. The concentration of sample 1(40% acetone precipitated) is 34.05 ug/mL and sample 2 (80% acetone precipitated) is 38.80 ug/mL.

Advantages

The biggest advantage is the speed of this method. The entire process takes about a half hour. This allows you to test several samples in a short amount of time. Using

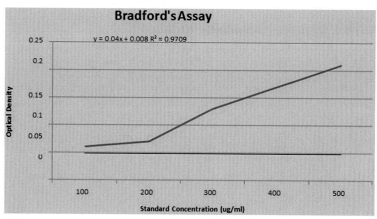

FIGURE 20.1

The concentration of protein is calculated by plotting the graph of standard protein.

Bradford can be advantageous against these molecules because they are compatible to each other and will not interfere. The test uses visible light (instead of UV light) to measure the absorbance of the sample. This way there is no need of a UV spectrophotometer but a simple visible light spectrophotometer. The Bradford assay is able to detect a large range of proteins, detecting amounts as small as 1 20 μg. It is an extremely sensitive technique and also very simple: measuring the OD at 595 nm after 5 min of incubation.

Limitations

The incompatibility with surfactants at concentrations is routinely used to solubilize membrane proteins. In general, the presence of a surfactant in the sample, even at low concentrations, causes precipitation of the reagent. The Bradford dye reagent is highly acidic, so proteins with poor acid solubility cannot be assayed with this reagent. Finally, Bradford reagents result in about twice as much protein-to-protein variation as copper chelation—based assay reagents. The Bradford assay is linear over a short range, typically from 0 μg/mL to 2000 μg/mL, often making dilutions of a sample necessary before analysis. In making these dilutions, error in one dilution is compounded in further dilutions resulting in a linear relationship that may not always be accurate.

Safety considerations and standards

The elevated concentrations of detergent like sodium dodecyl sulfate (SDS), a common detergent, may be found in protein extracts. This can cause underestimations of protein concentration in solution. Other interference may come from the buffer used when preparing the protein sample. A high concentration of buffer will cause an overestimated protein concentration due to depletion of free protons from the solution by conjugate base from the buffer. This will not be a problem if a low concentration of protein (subsequently the buffer) is used.

To identify a protein sample using SDS-PAGE

21

Chapter outline

Principle: Sodium dodecyl sulfate $(CH_3 (CH_2)_{10}CH_2OSO_3^- Na^+)$ SDS is a detergent that readily binds to the proteins. At pH 7 in the presence of 1%w/v SDS and 2-mercaptoethanol, proteins disassociate into their subunits and bind large quantities of detergent. Under these conditions, most proteins bind about 1.4 g of SDS per g of protein which completely masks the natural charge of proteins giving a constant charge to mass ratio. The larger the molecule, greater is the charge so the electrophoretic mobility of the complex depends on the size (mol. wt.) of the protein molecule and a plot of log of mol. wt. against relative mobility gives a straight line. In this scenario, the molecular wt of the protein is determined by comparing its mobility with series of protein standards. The sieving effect of the polyacrylamide is important in this technique and the range of molecular weights that can be separated on a particular gel depends on the pore size of the gel. The amount of cross-linking and hence pore size in a gel can be varied by simply altering the amount of acrylamide to make 5% or 10% gel.

Basic Life Science Methods. https://doi.org/10.1016/B978-0-443-19174-9.00021-0

Materials and equipment (materials and reagents)

- PAGE apparatus
- SDS buffer
- TEMED
- 40% acryl amide-(29:1) acrylamide:bis-acrylamide
- 1.5 M Tris-HCl (pH-8.8)
- 0.5 M Tris-HCl (pH-6.8)
- Sample buffer
- Ammonium persulfate
- Water bath
- 2-Mercaptoethanol
- Bromophenol blue
- Glycerol
- Coomassie blue
- Acetic acid
- Beakers
- Stirrer
- Measuring cylinder
- Agar
- **Reagent Preparation**
- **1.5 M Tris-HCl (pH 8.8) (for 10 mL)**

Dissolve 2.36 gm of Tris base in 5 mL distilled water. Maintain the pH to 8.8 and then make up the volume to 10 mL.

- **0.5 M Tris-HCl (pH 6.8) (for 10 mL)**

Dissolve 0.78 gm of Tris base in 5 mL distilled water. Maintain the pH to 6.8 and then make up the volume to 10 mL.

- **10% SDS buffer**

Dissolve 10 gm SDS in 20 mL distilled water and then make up the volume to 100 mL.

- **10% APS (freshly prepared)**

Dissolve 0.1 gm of APS in 1 mL of distilled water in an eppendorf.

- **TEMED (5% v/v)**
- **Acrylamide solution (40%) [29:1] (for 10 mL)**
- Acrylamide 3.857 gm
- Bis-acrylamide 0.133 gm
- Water 10 mL
- Filter and store the solution in a dark bottle.

Caution: the monomer is highly toxic, avoid inhalation and skin contact.

- **2X Sample loading dye for 1 mL**
  ddH$_2$O　125 µL
  0.5 M Tris, pH (6.8)　125 µL
  50% Glycerol　300 µL
  10% SDS　400 µL
  2-βmercaptoethanol　50 µL (add immediately before use)
  Pinch of Bromophenol Blue
- **1X Running Buffer for (1 L)**

  Tris—3.002 g.
  Glycine—14.40 g.
  SDS—1 g.

- **Staining Solution**

  Coomassie blue 0.2% w/v.
  Methanol: Acetic acid: water 5:1:5.
  Dissolve Coomassie Blue in Methanol: Acetic acid: Water.
  Make up the final volume to 500 mL.

Step-by-step method details (experimental procedure)
Procedure

1. Clean the surface thoroughly on which work has to be done.
2. Clean the glass plates with soap and water, then with ethanol. Assemble the glass plates and spacers.
3. Assemble the glass plates by putting the shorter glass plate in front of the longer one.
4. Fix the glass plates and fill it with distilled water to check any leakage from the bottom. If leakage occurs, seal the end of the plate with molten agar.
5. First, the resolving gel is prepared which is usually more basic and has higher polyacrylamide content than the loading gel.
6. Take a small beaker; add distilled water, Tris-Cal (pH-8.8), Acryl amide, SDS with the help of pipette.

Resolving gel (10%) [for 10 mL]

  H$_2$O　4 mL
  1.5 M Tris　2.5 mL
  10% SDS　100 µL
  40% Acrylamide　3.33 mL
  10% APS　50 µL
  TEMED　15 µL

Ammonium per sulfate and TEMED are added when the gel is ready to be polymerized.

Mix it properly by moving the beaker in circular motion on the surface.
Pour the gel between the plates before it polymerizes, then add water at the surface to remove unpolymerized gel.
Similarly prepare stacking gel using following composition.

Stacking gel (5%) [for 5 mL]

H_2O 3.075 ml
0.5M Tris   1.25 ml
10 % SDS   0.025 ml
40% Acrylamide   0.67 ml
10% APS   0.025 ml
TEMED   0.005 ml

When resolving gel polymerizes, remove the water by using tissue paper, pour the stacking gel and place the comb to create the wells.
After the stacking gel polymerizes, remove the comb and the gel is ready for electrophoresis.

Preparation of samples

➢ Mix the sample with an equal volume of 2X sample buffer. Boil the sample in boiling water for 5 min, cool to room temperature before loading. If particulate is present, centrifuge samples 2 min.

Loading the sample

➢ Clamp the gel plates properly and fill the buffer chamber with running buffer.
➢ Load 50 µg prepared sample into the wells. The tip is wiped with distilled water every time the next sample is to be loaded.
➢ First marker is loaded in the first lane to determine the size of desired protein.
➢ Then samples are loaded in the adjacent lanes. Bubbles are avoided in the tip.

Running the gel

➢ Place the lid on top of the buffer chamber.
➢ Connect the electrical leads to the power pack with the proper polarity (black to black and red to red) and run the gel at 60 V.
➢ Run the gel till the dye reaches the end of the gel.
➢ As soon as the dye reaches at the end, switch off the power.

Staining the gel

➢ Take out the glass plates from the electrophoretic tank and tear apart the two plates using spatula.
➢ Carefully place the gel into staining solution.

➢ Incubate the gel into this staining solution overnight.

Note: The staining of gel is always done in a covered container.

Destaining the gel

➢ Once the gel is stained, the gel is then transferred from a staining solution to a destaining solution.

➢ Destaining is done until all the extra dye is removed from the gel except the dye bound to protein in the form of bands.

➢ After destaining clear bands of protein are seen and compared to corresponding bands of ladder to determine the molecular weight of the protein.

Expected outcomes

Stacking and resolving gel were prepared by using distilled water 1.5 and 0.5M Tris Acryl amide, TEMED, APS. Protein sample was incubated at 95°C for 5 min along with gel loading dye. Samples were loaded in stacking gel. It was allowed to run through resolving gel. It was then stained with SDS-staining dye and destained overnight. Blue color bands show the presence of proteins.

Advantages

SDS polyacrylamide gel electrophoresis (SDS-PAGE) has the advantages of simple operation and good reproducibility in the determination of protein molecular weight, detection of specific proteins, and identification of strain species.

Limitations

An obvious limitation of SDS-PAGE resides in its deliberate denaturation of proteins prior to electrophoresis. Enzymatic activity, protein-binding interactions, detection of protein cofactors, etc., generally cannot be determined on proteins isolated by SDS-PAGE.

Safety considerations and standards

Wear a long-sleeved lab coat, safety goggles, nitrile gloves (latex is not effective), long pants, and closed-toe shoes. Wear appropriate skin and eye protection when working with UV radiation. Electrophoresis equipment can pose significant electrical hazards in the laboratory. Typical electrophoresis units operating at 100 V can provide a lethal shock of 25 milliamps. Take the following precautions when working with electrophoresis equipment:

- Ensure all switches and indicators are in proper working condition and that power cords and leads are undamaged and properly insulated.
- Turn off main power supply before connecting or disconnecting electrical leads.
- With dry gloved hands, connect one lead at a time using one hand only.
- Be sure that leads/banana plugs are fully seated.
- Measure, mix, and handle all hazardous powdered chemicals or gel prep mixtures with hazardous components (e.g., acrylamide monomer, ethidium bromide, phenol, ammonium persulfate, and formaldehyde) in the fume hood.
- Purchase premade gels or premixed acrylamide and ethidium bromide solutions instead of making your own.
- Exercise caution when using a microwave to melt agarose solutions—don't use sealed containers, and beware of superheated liquids that may suddenly and unexpectedly boil. Let hot agarose solutions cool to 50–60°C before adding ethidium bromide or pouring into trays. Wear insulated gloves and point the flask opening away from you.
- Gel chamber must have a lid or cover with safety interlocks to prevent accidental contact with energized electrodes or buffer solutions.
- Gel chamber exterior must be dry with no spilled solutions. Check the chamber for leaks.
- Switch off all power supplies and unplug the leads before opening the gel chamber lid or reaching inside the gel chamber. Do not rely on safety interlocks.

Genomic DNA isolation from whole blood

Chapter outline

Principle

Deoxyribonucleic acid or DNA is a genetic material, located in the nucleus of the cell and in the cellular organelle like mitochondria. The salting-out method is a simple and nontoxic DNA extraction technique, introduced by Miller et al. that isolates a high-quality DNA from the whole blood (Miller et al., 1988). In the standard salting-out method, proteins K and RNase are added to them after the lysis of cells. Saturated NaCl was needed for the proteins to precipitate out of the solution. Genomic DNA was isolated from blood using salting out method, as described below.

Basic Life Science Methods. https://doi.org/10.1016/B978-0-443-19174-9.00022-2

Materials and equipment (materials and reagents) (Tables 22.1–22.3)

- Tris-HCl
- Potassium chloride
- Magnesium chloride
- EDTA
- Sodium chloride
- Sodium dodecyl sulfate
- Isopropanol
- Ethanol
- Triton-X
- 1.5 mL eppendorf tubes
- centrifuge (REMI, research centrifuge)
- **Preparation of Reagents:**

    The reagents were prepared as described below:

    a. TKM1 buffer/low salt buffer (500 mL):

Table 22.1 Composition of TKM1 buffer.

Chemical	Amount
Tris HCl (10 mM) pH 7.6	0.605 g
KCl (10 mM)	0.372 g
MgCl$_2$ (10 mM)	1.016 g
EDTA (2 mM)	1.372

    b. *Triton-X (10 mL):* Added 0.1 mL of 100% Triton-X to 9.9 mL of distilled water.

    c. **TKM2 Buffer/High salt buffer: (100 mL)**

Table 22.2 Composition of TKM2 buffer.

Chemical	Amount
Tris HCl (10 mM) pH 7.6	0.121 g
KCl (10 mM)	0.074 g
MgCl$_2$ (10 mM)	1.203 g
EDTA (2 mM)	0.074 g
NaCl (0.4 M)	1.467

d. *SDS (10%) (10 mL):* Dissolved 1 g of sodium dodecyl sulfate in distilled water.
e. *6M NaCl (25 mL):* Dissolved 8.765 g of NaCl in 25 mL of distilled water.
f. **Tris EDTA (TE) buffer (25 mL):**

Table 22.3 Composition of tris EDTA.

Chemical	Amount
Tris HCl (10 mM) pH 8.0	0.030 g
EDTA (1 mM)	0.009 g

Step-by-step method details (experimental procedure)
Procedure
RBC lysis

1. 300 μL of blood is treated with 900 μL TKM1 and 90 μL of 1x Triton-X for 5 minutes at room temperature (RT).
2. After vortexing, the solution is centrifuged at 8000 rpm for 5 minute and the supernatant is discarded after centrifugation.
3. This same step is repeated for minimum 3 times to get a white pellet of WBCs.

Cell lysis

1. After the RBC lysis step pellet is treated with 260 μL of cell lysis buffer (TKM2) and 40 μL of the 10% SDS and pipetted thoroughly to lyse the cells.
2. Mixed thoroughly and incubated at room temperature for 15 minutes.
3. At the end of the incubation, 100 μL of 6M NaCl is added and vortexed to precipitate the proteins.
4. At the end of the incubation, 100 μL of 6M NaCl is added and vortexed to precipitate the proteins.
5. Cells are centrifuged at 8000 rpm for 5 minutes.

Precipitation of DNA

1. The supernatant is transferred into a new eppendorf tube containing 300 μL of isopropanol. DNA is precipitated by inverting the eppendorf slowly.
2. Further, the eppendorfs are centrifuged at 8000 rpm for 10 minutes to pellet down the DNA.
3. Supernatant is discarded, 70% ethanol added and mixed slowly to remove any excess salts.
4. Finally, the tubes are centrifuged at 8000 rpm for 5 minutes to pellet down the DNA.
5. Supernatant is discarded and pellet is air-dried.
6. After air drying, 50–100 μL of TE buffer is added to dissolve the DNA.

DNA quantification

DNA is quantified on nano-drop to check the quality. The good quality DNA is stored at −20°C, until further use.

Expected outcomes

The quality and quantity of extracted DNA can be measured spectrophotometrically at wave length 260 and 280 nm. A good quality DNA extracted from whole blood is shown in Fig. 22.1.

Advantages

The main advantage of this method is simple, fast, and inexpensive method, which can be used for genetic analysis in medical laboratories and research centers with financial and time limitations. In addition, our results indicate that the DNA produced by this modified method with high quality could be used for PCR-based experiments, especially for gene polymorphism studies in a human population.

Limitations

It is a time-consuming process. The use of proteinase K can be time-consuming and expensive when compared with other reagents used in different solution-based approaches.

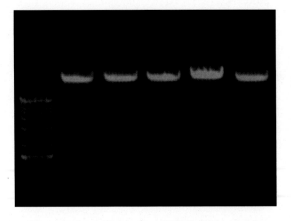

FIGURE 22.1

Representative gel picture of DNA extracted from blood.

Safety considerations and standards

(a) Correct handling and storage of starting material. (b) Perform extractions at 4°C, on ice or in the cold. (c) Inhibit nuclease activity. (d) Store purified DNA correctly.

Reference

Miller, S.A., Dykes, D.D., Polesky, H.F., 1988. A simple salting out procedure for extracting DNA from human nucleated cells. Nucleic Acids Research 16 (3), 1215.

PCR for gene amplification

23

Chapter outline

Principle: The PCR or polymerase chain reaction is a simple and rapid technique in molecular biology for duplicating a DNA sequence in many copies. Polymerase enzymes, which aid in DNA replication, are used to facilitate this method of DNA synthesis by adding nucleotides to a single stranded template at $3'$ end using primers. Because of its sensitivity and specificity, PCR has quickly become a standard molecular biology technology with a wide range of applications. The double-stranded DNA is first denatured for a short period which breaks the weak hydrogen bonds then the split strands may anneal with the primers (forward and reverse). The two strands are then allowed to lengthen utilizing the deoxynucleoside triphosphate in the PCR mix, resulting in new strand stretches that are complementary to the template strands. Denaturation, annealing, and extension cycles are carried out in around 30–35 cycles in thermo cycler during which the original sequence of DNA is amplified exponentially.

Materials and equipment

- 10x PCR buffer
- 25 mM $MgCl_2$
- 10 mM dNTPs (Eppendorf)
- Autoclaved double distilled water (ddH$_2$O)
- Primers (Forward and Reverse)

Basic Life Science Methods. https://doi.org/10.1016/B978-0-443-19174-9.00023-4

- Taq DNA polymerase (5U/uL)
- DNA template
- 10% DMSO
- Thermal cycler
- PCR tubes and pipettes.

Procedure

1. For making a 50 uL reaction (Table 23.1)
2. Take a PCR eppendorf tubes and label it.
3. Pipette into the PCR tube as shown in table below.
4. Mix well and keep it in PCR thermal cycler.
5. After mixing all the contents of the reaction, PCR tube should be placed in thermal cycler and a three-step PCR with an initial denaturation at 94°C for 5 min followed by cycling at 94°C for 30 s, annealing for 30 s–1 min at appropriate temperature, 72°C for 45 s and a final extension at 72°C for 5 min was carried out for about 35 cycles.

Table 23.1 PCR reaction mix.

Chemical	Amount
Distilled water	36.5 µL
10x buffer	5 µL
MgCl$_2$	4 µL (2 mM)
dNTPs	1 µL (10 mM/uL)
Forward primer	1 µL (10 pmol/uL)
Reverse primer	1 µL (10 pmol/uL)
Template DNA	1 µL
Taq Polymerase	0.5 µL (2.5 U)

Below is a list of characteristics that should be considered when designing primers.

1. The length of the primer should be 15–25 nucleotide residues (bases).
2. The GC content of primer should range between 40% and 60%.
3. G or C should be added to the 3′ end of primers to clamp the primer and prevent "breathing" of the ends, enhancing priming efficiency. When the ends of DNA do not stay annealed but tear or split apart, this is known as "breathing." The three hydrogen bonds in GC pairings assist inhibit breathing while also raising the primers' melting temperature.
4. The 3′ extremities of a primer set that comprises a plus and minus strand primer should never be complimentary to one other, and the 3′ end of a single primer

should not be complementary to other sequences in the primer. In these two circumstances, primer dimers and hairpin loop structures are formed, respectively.

5. Primer melting temperatures (Tm) are best between 52 and 58°C, however the range can be extended to 45−65°C. The difference in final Tm between the two primers should be no more than 5°C.

6. Single base runs (e.g., AAAAA or CCCCC) or di-nucleotide repetitions (e.g., GCGCGCGCGC or ATATATATAT) should be avoided since they might induce slippage down the primed segment of DNA and/or the formation of hairpin loop structures. Only include repetitions or single base runs with a maximum of four bases if it is unavoidable owing to the nature of the DNA template.

Notes

There are many tools to design primer pairs
http://www.ncbi.nlm.nih.gov/tools/primer-blast/
Primer3 http://frodo.wi.mit.edu/primer3.

Expected outcomes

A PCR reaction's purpose is usually to duplicate only a small section of the genome of interest. As a result, a new strand of DNA and a double-stranded DNA molecule are synthesized. The length of the DNA sequence to be amplified depends upon the duration of about 1 min for duplication of 1000 DNA bases.

Advantages

The benefits of PCR are numerous. It's a method that is becoming increasingly important in molecular biology and related subjects. PCR is commonly used in diagnostics and to identify the presence of a certain DNA sequence of any organism in a biological fluid. It provides an alternative method for cloning. PCR is used to determine genotyping, DNA methylation, and phylogenetic analysis.

Limitations

(a) PCR can't be utilized to amplify targets that aren't known. To construct the primers, prior knowledge about the target sequence is required.

(b) DNA polymerases are prone to errors, which might result in PCR product mutations.

(c) PCR is extremely vulnerable to contamination.

Safety considerations and standards

(a) The PCR method produces a huge number of amplicons. As a result, it poses a significant danger of contamination. Even minute quantities of contaminating DNA can result in large amplicon concentrations and erroneous findings.

(b) Decontaminating and cleaning the lab environment.

(c) Always use gloves and a lab coat when doing operations.

Agarose gel electrophoresis

Chapter outline

Principle: Agarose is a polysaccharide, generally extracted from certain red seaweed. It is a linear polymer made up of the repeating unit of agarobiose, which is a disaccharide made up of D-galactose and 3,6-anhydro-L-galactopyranose. Fragments of linear DNA migrate through agarose gels with a mobility that is inversely proportional to the log 10 of their molecular weight. By using gels with different concentrations of agarose, one can resolve different sizes of DNA fragments. Higher concentrations of agarose facilitate separation of small DNAs, while low agarose concentrations allow resolution of larger DNAs.

Materials and equipment (materials and reagents)

- Electrophoresis chamber and power supply
- Gel casting trays
- Combs
- Electrophoresis buffer (TBE)
- Loading dye
- Ethidium bromide (10 mg/mL)
- Agarose (Sigma)
- 100 bp or 50 DNA marker
- Electronic weighing balance

- Microwave oven
- Gel documentation system

Preparation of 100 mL 10x TBE (Table 24.1).

Table 24.1 TBE buffer composition.

Chemical	Amount
Tris base	10.3 g
Boric acid	5.4 g
EDTA (0.5 M)	4 mL
H_2O	100 mL

Preparation of 6x loading dye: (Table 24.2).

Table 24.2 Loading dye composition.

Chemical	Amount
Xylene cyanol	0.25%
Bromo phenol	0.25%
Glycerol	4 mL
H_2O	100 mL

Step-by-step method details (experimental procedure)
Procedure

1. Preparation of 2% agarose gel:
   - One gram of agarose should be added to 50 mL of 1x TBE buffer in a 100 mL conical flask or 2 gm in 100 mL.
   - The flask should be kept in a microoven and boiled until the agarose dissolved.
   - The solution should be allowed to cool and 2 μL of 10 ng/mL ethidium bromide was added.
   - Mixed well and pour into the gel-casting tray.
   - The comb should be kept in place and the gel should be allowed to solidify at room temperature.
2. Sample loading and electrophoresis:
   - After 20–30 min, the comb should be removed and the gel must be placed in the electrophoresis chamber.

- 10 μL of the PCR product should be mixed with 1 μL of 6x loading dye and loaded carefully in to the wells.
- 2−4 μL of 100 bp DNA ladder is added to one of the wells.
- Samples are run at 150 V for 45 min.

3. Image capturing and molecular weight analysis:
- After the run, the gel should be removed from the casting tray and slided on to the transilluminator (Fig. 24.1).
- Switched on UV and capture the picture using the gel documentation system.
- Finally, the picture is transferred to the computer and the molecular weight is determined using the UVI tech software.

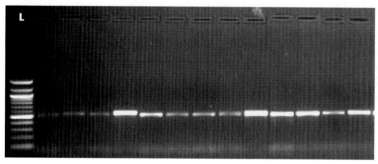

FIGURE 24.1

Representative gel picture of PCR product run on 2% gel.

Expected outcomes

Agarose gel electrophoresis is used to know how many different DNA fragments are present in a sample and how large they are relative to one another. By comparing the bands in a sample to the DNA ladder, we can determine their approximate sizes.

Advantages

The main advantage of agarose gel technique is that it can be easily processed and the DNA molecule that is used as a sample can also be recovered without any harm to it at the end of the process. Agarose gel does not denature the DNA sample and they stay in their own form.

Limitations

(a) It melts when an electric current is passed through it. Due to this reason, there are chances that genetic material can adopt the shapes which are not needed.

(b) Electrophoresis has limited sample analysis. (c) Electrophoresis is specific to whatever tissue you've sampled. (d) Electrophoresis Measurements Are Not Precise. (e) Substantial Starting Sample is Required.

Safety considerations and standards

1. When locating, or working around or near an electrophoresis unit, avoid unintentional grounding points and conductors (e.g., sinks and other water sources, metal plates, aluminum foil, jewelry, pipes, or other metal equipment).
2. Always think and look before touching any part of the apparatus. A thin film of moisture can act as a good conductor of electricity.
3. Some power supplies produce high voltage surges when they are first turned on, even if the voltage is set to zero. Do not ignore safety rules just because the voltage is low. Changes in load, equipment failure, or power surges could raise the voltage at any time.
4. Do not run electrophoresis equipment while unattended.
5. If electrophoresis buffer is spilled or leaks from the gel box, stop the run and clean up the bench top immediately.
6. Post "Danger—High Voltage" warning signs on the power supply and buffer tanks
7. Principal Investigators or Lab Managers are responsible for providing instruction and demonstrating safe use of electrophoresis units to laboratory workers. Instruction should cover operating procedures written by the manufacturer and/or laboratory, as well as the associated hazards, the correct personal protective equipment (lab coats, gloves, and eye protection), and applicable emergency.

RNA isolation from human tissue* 25

Chapter outline

RNA is quickly destroyed by the ubiquitous RNases. As a result, all tubes and solutions used in this approach must be RNAse-free (simple autoclaving does NOT inactivate RNases). When dealing with RNA, the need of a clean work environment cannot be overstated.

Some important precautions

Because of the presence of difficult-to-inactivate RNases on human skin and in bodily fluids, RNA in the lab is exceedingly labile. These modest measures, on the other hand, can help to reduce the degradation of valuable RNA samples.

1. Put on a clean pair of gloves and a lab coat. Avoid touching your face or hair with gloved hands. Remove gloves before contacting objects that are normally touched with bare hands, such as a computer. Frequently change your gloves.
2. Before beginning work, clean the workspace and pipettes with RNaseZap.
3. For RNA work, use RNase-free barrier pipette tips and microfuge tubes. These do not need to be autoclaved if they are RNase-free.
4. All solutions should be prepared using RNase-free water and solely used for RNA processing. RNase-free water can be purchased commercially or produced

---

* Note: Handling Conditions Standard Precautions must be followed when handling all solid tissue samples. Samples can be stored in RNA later followed by storage in a $-80°C$ freezer.

Basic Life Science Methods. https://doi.org/10.1016/B978-0-443-19174-9.00025-8

via DEPC treatment of ultrapure water. Add 0.1% v/v DEPC (diethylpyrocarbonate; very hazardous) to water and mix overnight before autoclaving.

5. Use proper sanitary technique: if feasible, keep tubes and bottles covered, and avoid coughing, sneezing, or breathing over open containers.
6. Keep RNA-containing samples on ice whenever possible, and purified RNA in a separate box at $-80°C$.

Requirements

Centrifuge, Magnetic Stir Plate, Microcentrifuge, Homogenizer, Nanodrop, Vortex Mixer, Water Bath, Microcentrifuge Tubes, Chloroform, Nuclease Free Water, 75% Ethanol ($-20°C$), Isopropanol, TRIzol reagent, DNA-free Kit, DNAse I buffer, DNase, Inactivation reagent.

Protocol

1. Frozen tissue should be kept immersed in dry ice until it is mixed with Trizol.
2. Pipet the homogenized material into a fresh 2 mL screw cap tube with 0.5 mL of TRIzol. To mix, invert the tube. Place the tube on wet ice or dry ice until ready to go to Phase Separation.
3. Incubate at room temperature for 5 min.
4. Shake gently by hand for 15 s after adding 0.2 mL of chloroform to 1 mL of TRIzol reagent. Incubate the samples for 3 min at room temperature.
5. Centrifuge the samples for 15 min at $4°C$ at $11,600 \times g$.
6. The mixture separates into a colorless, higher aqueous phase and a lower red phenol-chloroform phase. RNA is only found in the aqueous phase (which is about 60% of the volume of TRIzol reagent used for initial homogenization). The organic phase, on the other hand, can be preserved for later DNA and protein extraction.
7. Add 0.5 mL of isopropanol.
8. Transfer the aqueous phase to the isopropanol tube that has been labeled. Pipette up and down gently to remove the RNA from the aqueous phase. (Avoid vortexing.)
9. After 10 min at room temperature, centrifuge the samples at $11,600 \times g$ for 10 min at $4°C$.
10. The RNA precipitate forms a transparent gel-like pellet on the tube's side and bottom.
11. Discard the supernatant and wash the RNA pellet once with ice-cold ethanol (75%).
12. Centrifuge for 5 min at $4°C$ at $7500 \times g$. (Avoid Vortexing)

13. Dissolve the pellet in 20–50 L of RNase-free water and store in a −80°C freezer.

Dilute sample in RNase-free water, then measure absorbance at 260 and 280 nm. Calculate the RNA concentration using the formula A260 × dilution × 40 = µg RNA/mL.

Calculate the A260/A280 ratio.

A ratio of ~2 is considered pure.

Optimized protocol for the extraction of RNA isolation from blood

Venous blood samples (5 mL) should be taken from participants and placed in EDTA-Na2 tubes. The filled EDTA tubes should be put in a 20°C freezer right away and then transferred to an 80°C deep freezer the next day. The frozen samples should be thawed in the following ways: in a 37°C water bath for about 5 min, on an aluminum block at room temperature (23°C) for about 18 min, or on ice for 2 h. When compared to thawing the samples in a 37°C water bath, quickly thawing frozen whole blood on aluminum blocks at room temperature might reduce RNA degradation and enhance RNA production and quality. Furthermore, as compared to the PAXgene Blood RNA Kit, the NucleoSpin RNA kit boosted RNA production by fivefold. Thawing blood samples on aluminum blocks boosted DNA production by 20% when compared to thawing in a 37°C water bath or on ice. Furthermore, by thawing frozen EDTA whole blood samples on aluminum blocks and using the NucleoSpin RNA and QIAamp DNA Blood kits, adequate quality and quantity of RNA and DNA can be extracted from frozen EDTA whole blood samples preserved for longer period. Thus, it is possible to recover RNA from frozen whole blood in EDTA tubes after long-term preservation (Yamagata et al., 2021).

Reference

Yamagata, H., Kobayashi, A., Tsunedomi, R., Seki, T., Kobayashi, M., Hagiwara, K., Nakagawa, S., 2021. Optimized protocol for the extraction of RNA and DNA from frozen whole blood sample stored in a single EDTA tube. Scientific Reports 11 (1), 1–10.

Real-time polymerase chain reaction

26

Chapter outline

Principle: Real-Time PCR is a special technique that allows you to see the development of a PCR reaction "in real-time." The fluorescence produced by a reporter molecule, such as Taq man Probe or Sybr green dye, is detected in real-time PCR. This happens because the PCR product multiplies with each amplification cycle (Fig. 26.1). Dyes that bind to double-stranded DNA are included in these fluorescent reporter molecules. The number of cycles necessary for the fluorescent signal to reach the threshold is known as the Ct (cycle threshold) (i.e., exceeds background level).

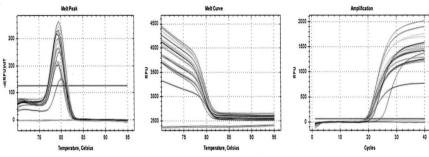

FIGURE 26.1

Melting curve analysis is a method of determining how double-stranded DNA dissociates during heating. The double strand begins to dissolve as the temperature rises, resulting in an increase in absorbance intensity, or hyperchromicity. Melting temperature is defined as the temperature at which 50% of DNA is denatured. Real-time PCR amplification plot showing the accumulation of product over the duration of the real-time PCR experiment.

Types of real-time quantification
Absolute quantitation

To produce a standard curve in absolute quantitation, serially diluted standards of known concentrations are used. The standard curve establishes a linear connection between Ct and beginning levels of total RNA or cDNA, allowing the concentration of unknowns to be determined using Ct values. This approach implies that all standards and samples have amplification efficiencies that are about identical. Furthermore, the concentration of serial dilutions should match the levels in the experimental samples while remaining within the range of reliably measurable and detectable levels for both the real-time PCR equipment and the assay. DNA standards have been found to have a wider measurement range, as well as higher sensitivity, repeatability, and stability than RNA standards. Due to the lack of a reverse transcription efficiency control, a DNA standard cannot be employed for a one-step real-time RT-PCR (Souazé et al., 1996; Fronhoffs et al., 2002).

Relative quantitation

Changes in sample gene expression are quantified using either an external standard or a reference sample, commonly known as a calibrator, during relative quantification. The results of utilizing a calibrator are expressed as a target/reference ratio. The mean normalized gene expression from relative quantitation tests may be calculated using a variety of statistical approaches. Depending on the approach used, they might produce different findings and consequently differing standard error measurements (Muller et al., 2002).

Materials and equipment

- Reagents:
- Random hexamer (50 ng/μL)
- Reverse transcriptase MMLV-RT (50 units/μL)
- Bio-Rad thermo cycler
- Ethidium bromide
- Gel Doc
- 2% agarose gel

- 10× TaqMan RT Buffer
- dNTP (2.5 mM each dNTP)
- RNase inhibitor (20 units/μL)
- TriZol
- Chloroform
- Isopropanol

- RNeasy mini kit method (QIAGEN)
- 1 μL of cDNA, 12.5 μL eva green
- Forward primer 5 μM
- Reverse primer 5 μM
- Centrifuge
- 50× TAE Buffer
- 70% ethanol

Step-by-step method details
Procedure
Extraction of total RNA

1. Total RNA is isolated from tissue sample, using the RNeasy Mini Kit method (QIAGEN) or manual method.
2. The concentration and purity of extracted RNA are determined by measuring the absorbance at 260/280 nm.
3. All samples whose concentration is a minimum of 100 ng/μL at 260/280 between 1.8 and 2 are included in the study.
4. Alternately the samples are run on agarose to check for the integrity.

Reverse transcription
Reverse transcription was performed as per manufacturer protocol

First strand cDNA is synthesized from isolated RNA using oligo-dT and superscript II (Invitrogen) in a reaction volume of total 20 μL. cDNA synthesis is done using First Strand cDNA Synthesis Kit using oligodT primers with some modifications. The first strand cDNA is synthesized by using manufacturer's protocol as follows. The cDNA synthesized is stored at −20°C. The protocol for synthesis of first strand is given below.

- The following reagents are added in an RNase-free tube on ice as per the indicated order.

Template RNA	1–2 μL
Oligo(dT)$_{18}$ primer	1 μL
5× buffer	4 μL
Nuclease free water	7–8 μL

- The components were mixed gently and a brief spin is given.

- Incubation at 65°C for 5 min is given.
- The tubes are chilled on ice and brief spin is given.
- The tubes are again placed on ice and the following components are added as per the indicated order.

RiboLock RNase Inhibitor (20 U/μL)	1 μL
10 mM dNTP mix	2 μL
RevertAid M-MuLV RT (200U/μL)	2 μL

- The components are mixed gently and a brief spin are given.
- The components are incubated at 42°C for 40 min.
- The reaction is terminated by heating at 70°C for 5 min.

Real-time PCR

The PCR reaction mixture for relative gene expression study is given below.

Constituents	Volume
SYBR green supermix (2×)	5 μL
Primer mix (10 μM)	1 μL
cDNA template	1 μL
Nuclease free water	3 μL
Total volume	**10 μL**

Relative gene expression studies

The relative gene expression is performed according to manufacturer's instructions (BIORAD, Californian, USA). The PCR conditions are specified as (95°C for 3 min, 40 cycles each at 95°C for 30 s and 50–62°C for 30 s) or can vary according to requirement. The reference gene to be used can be actin/GAPDH to calculate relative gene expression of candidate genes. All the reactions are carried out in three technical triplicates and no-template controls (NTC) (Mansoor et al., 2020). The quantification of target pathway gene to be determined relative to actin/GAPDH using the $\Delta\Delta Ct$ method and expressed as $2-\Delta\Delta\ CT$ for graphic representation. Relative gene expressions will be determined according to the Livak and Schmittgen (2001):

Relative Gene Expression $= 2^{-\Delta\Delta Ct}$, where $\Delta\Delta C_t = [\Delta]\ Ct_{sample} - [\Delta]\ Ct_{reference}$

$[\Delta]\ C_{t\ sample}$—C_t value for any sample normalized to the endogenous housekeeping gene

$[\Delta]\ C_{t\ reference}$—$C_t$ value for the reference sample normalized to the endogenous housekeeping gene

Expected outcomes

Tissue-specific gene expression is determined using real-time PCR. The current technique has been used to track cancer patients' prognosis and treatment responses. It determines how many tissue-specific mutant alleles there are. It has a lot of uses in gene insertion investigations and gene therapy trials. Its innovative technique also allows for illness detection and microbiological identification. The viral load, expression, and infection may all be measured by RT-PCR.

Advantages

Real-time qPCR has a plethora of applications in several domains, ranging from gene measurement to gene expression. It is also utilized in the food industry, microbial identification, and illness diagnostics, in addition to gene expression investigations.

Limitations

When compared to traditional PCR, the instrument is too expensive. Because the procedure is so sensitive, even a trace of DNA contamination might cause erroneous findings. Only a few tests make advantage of this approach. To execute and create novel tests, a great deal of knowledge and competence is necessary.

Safety considerations and standards

To avoid RNase contamination, always use gloves. RNase Inhibitor and Reverse Transcriptase should be placed on ice right out of the box. Wipe out workstations with bleach or commercially available decontamination treatments in dilute concentrations. Samples should be prepared in a dedicated clean room, hood, or benchtop workstation with a UV lamp. This should ideally be in a separate area from the heat cycler. During sample preparation and reaction setup, change gloves often. Use aerosol-barrier pipet tips and shaft guard pipets. Pipets should be cleaned with a weak bleach solution on a regular basis. Use PCR-grade water and reagents that are only for PCR. For dilutions and reaction preparation, use screw-capped tubes.

References

Fronhoffs, S., Totzke, G., Stier, S., Wernert, N., Rothe, M., Brüning, T., Ko, Y., 2002. A method for the rapid construction of cRNA standard curves in quantitative real-time reverse transcription polymerase chain reaction. Molecular and Cellular Probes 16 (2), 99–110.

Livak, K.J., Schmittgen, T.D., 2001. Analysis of relative gene expression data using real-time quantitative PCR and the 2−ΔΔCT method. Methods 25 (4), 402−408.

Mansoor, S., Sharma, V., Mir, M.A., Mir, J.I., un Nabi, S., Ahmed, N., Masoodi, K.Z., 2020. Quantification of polyphenolic compounds and relative gene expression studies of phenyl-propanoid pathway in apple (*Malus domestica* Borkh) in response to Venturia inaequalis infection. Saudi Journal of Biological Sciences 27 (12), 3397−3404.

Muller, P.Y., Janovjak, H., Miserez, A.R., Dobbie, Z., 2002. Short technical report processing of gene expression data generated by quantitative real-time RT-PCR. Biotechniques 32 (6), 1372−1379.

Souazé, F., Ntodou-Thome, A., Tran, C.Y., Rostene, W., Forgez, P., 1996. Quantitative RT-PCR: limits and accuracy. Biotechniques 21 (2), 280−285.

Preparation of competent cells by CaCl₂ treatment

27

Chapter outline

Principle

As DNA is highly hydrophilic molecule, normally it cannot pass through the cell membrane of bacteria. Hence, in order to make bacteria capable of internalizing the genetic material, they must be made competent to take up the DNA. This can be achieved by making small holes in bacterial cells by suspending them in a solution containing a high concentration of calcium. Extra chromosomal DNA will be forced to enter the cell by incubating the competent cells and the DNA together on ice followed by a brief heat shock that causes the bacteria to take up the DNA.

Materials required

Glassware: Conical Flasks, Falcon Tube, Eppendorf Tubes, Micropipettes, Sterile Petri Plates, Ice Buckets

    Reagent Required: LB Broth LB Agar 0.1 M CaCl, Host Cell Stock, Ampicillin (100 mg/mL).

Basic Life Science Methods. https://doi.org/10.1016/B978-0-443-19174-9.00027-1

Equipment

Autoclave, Spirit Lamp, Ice Flaker, BOD Incubator Shaker, Refrigerator, Deep Freezer [−20 to −80°C], Vortex.

Principle

Competent cells are bacterial cells that can expect extrachromosomal DNA or plasmids from the environment. The generation of competent cells may occur by two methods—natural competence or artificial competence.

Natural competence is the genetic ability of a bacterium to receive environmental DNA under natural or in vitro conditions. Bacteria can also be made competent artificially by chemical treatment and heat shock to make them transiently permeable to DNA.

Preparation of media or solution

1. Prepare LB Agar for petri plates:
   Weigh 2 gm of LB Broth and 2 gm of Agar and make final volume 100 mL for 5 petri plates in a conical flask. Autoclave the solution and after that add 100 ug/mL ampicillin to final concentration. [100 uL from 100 mg/mL stock solution to 100 mL of LB Agar media solution]. Now pour 20 mL to each plate and let them cool down to room temperature. After they get solidified, store them at 4−8°C in a refrigerator.
2. Prepare LB liquid media:
   Add 4 gm of LB broth in a conical flask and add 200 mL of water. Autoclave the solution.
3. Prepare 0.1 M $CaCl_2$:
   Weigh 1.47 g of $CaCl_2$ in a beaker and make final volume to 100 mL. Autoclave the solution.

Procedure

Overnight culture

- We take two falcon tubes—Pure culture and control.
- In pure culture we add 15 mL of LB Broth and 40 uL of overnight culture.
- In control we add 15 mL of LB Broth and 40 uL of overnight culture. Also, we add 7 uL of Ampicillin. We make two tubes to ensure that our culture contains any resistance genes or not.
- Place the tubes in shaking incubator at 37°C at 200 rpm.

- Incubate for 12–16 h.
- Next day we saw that pure culture shows growth while control shows no growth.
- So, our culture does not contain any resistance genes.

Generation of competent cells

- Add 5 mL of pure culture to 100 mL of fresh LB Broth [if pure culture shows less growth, we can use more] in conical flask.
- Place the flask in incubator shaker at 37°C at 200 rpm for 3–4 h or until OD reaches 0.4–0.6 [at this OD, cells will be in log phase].

CaCl$_2$ wash

- Ensure all solutions (CaCl$_2$ tubes) are ice cold or at 4°C.
- Now separate this culture into various falcon tubes [usually we take 2 each of 50 mL].
- Centrifuge the tube for 5 min at 900 rpm and at 4°C.
- Discard the supernatant and wash with some CaCl$_2$ so that media is removed out.
- Now add 20–30 uL of CaCl$_2$ to pellet and mix with cells with the help of vortex.
- Now incubate on ice for 30–40 min.
- Again, centrifuge the tube at 9000 rpm for 5 min at 4°C temperature.
- Discard the supernatant and add 1 mL CaCl$_2$ and also add 10% glycerol.
- Now the cells can be stored at—80°C for a week or used directly for transformation.

  Experiment: Preparation of calcium competent *Escherichia coli*.

Materials and equipment

A. Competent cell preparation:

- 1 mL of overnight *Escherichia coli* culture
- 100 mL of 0.1 M CaCl$_2$ (ice cold)
- 20 mL of 0.1 M CaCl$_2$ with 15% glycerol solution (ice cold)
- 100 mL fresh lysogeny broth (LB) media

B. Equipment:

- 37°C shaking incubator
- 42°C water bath
- Spectrophotometer

Methods

1. CaCl$_2$ buffer preparation

- 1 M CaCl$_2$ (working solution, 10× working concentration)
- Weight out 11.1 g of anhydrous CaCl$_2$
- Add to 80 mL of distilled water

- Mix solution until $CaCl_2$ is fully dissolved
- Top up to 100 mL
- Filter sterilizes through a 0.22 um pore

2. 0.1 M $CaCl_2$ (working solution)
   - Add 10 mL of 1M $CaCl_2$ to 90 mL of distilled water for a 1:10 dilutions
   - Filter sterilizes through a 0.22 um pore

3. 0.1 M $CaCl_2$ + 15% glycerol (working solution)
   - Mix 6 mL 1 M 0.1 M $CaCl_2$ with 9 mL sterile glycerol and 45 mL distilled water
     ➤ Overnight Culture(s)
   - Inoculate 1 mL of LB with *E. coli*
   - Place in shaking incubator at 37°C and 200 rpm
   - Incubate for 12−16 h
     ➤ Generation of competent cells ($CaCl_2$ wash)
   - Subculturing overnight culture:
   - Add 1 mL of overnight culture to 99 mL of fresh LB (1:100 Dilution, no antibiotics)
   - Shake incubate at 37°C and 200 rpm for 3−4 h or until OD reaches 0.4
     ➤ $CaCl_2$ wash:
   - Ensure that all reagents (**$CaCl_2$ solutions, Oakridge tubes**)
   - Place on ice for 20 min
   - Centrifuge at 4°C at 4000 rpm for 10 min
   - Discard the supernatant by tipping tubes over a discard bin and then aspirating any remaining media
   - Resuspend each pellet with 20 mL ice cold 0.1 M $CaCl_2$ with 15% glycerol
   - Use for downstream transformation or store in −80°C freezer

Heat shock transformation of *Escherichia coli*
Materials and equipment

A. Competent cell preparation:
   - 1 mL of competent cells of *Escherichia coli*
   - Ice cold
   - 100 mL fresh broth (LB) media

B. Equipment:
   - 37°C shaking incubator
   - 42°C water bath

Methods
Heat shock

➤ Thaw competent cells on ice:
   - Add 1−5 uL (10−100 ng) of plasmid (do not exceed 5 uL for 50 uL cell aliquot).

- Incubate on ice for 30 min.
- Heat shock by placing in 42°C water bath for exactly 30 s.
- Place cells on ice for 2 min.
- Add 1 mL prewarmed LB or SOC medium.
- Shake and incubate at 37°C, 200 rpm, 1 h for outgrowth.

➤ Slanting and incubation:
- Spread plate 1:10 and 1:100 dilutions of the outgrowth cultures on warm selective and/or screening plates (e.g., Ampicillin and/or X-gal if required).
- Incubate at 37°C for 12−16 h.

➤ Plate observations:

Inspect plates for isolated colonies.

➤ Controls:
- Use a DNA preparation that has been shown to give transformants in previous experiments to act as a positive control.
- Perform another transformation to which plasmid DNA is not added to act as a negative control.
- Transform 1 ng of target plasmid to check competent cell viability, calculate transformation efficiency, and verify the antibiotic resistance of the plasmid.

RNA isolation from plant tissues

Chapter outline

Prior to RNA isolation, microfuge tubes, tips, tip boxes, mortar-pestle, spatula, and forceps should be DEPC treated (0.1%) overnight at 37°C and then autoclaved followed by drying in hot air oven. Extraction buffer—2% CTAB, 2% polyvinylpyrrolidone (PVP) K-30 (soluble), 100 mM Tris HCl (pH 8.0), 25 mM EDTA, 2.0 M NaCl, 0.5 g/L spermidine (free acid) (HRS), 2% β-mercaptoethanol (added just before use) with several modifications to extract RNA fast and method is as follows (Asif et al., 2006):

➢ Tissue samples (100 mg) are grinded in a pestle-mortar and in liquid nitrogen and transferred to a fresh microfuge tube.
➢ Take prewarmed CTAB buffer (at 60°C), 200 μL of 2% β-mercaptoethanol is added to the homogenized sample and mixed thoroughly and incubated at 60°C for more than half an hour and vortexed after every 10 min.
➢ Equal volumes of chloroform:isoamyl alcohol (24:1) is added and immediately vortexed for 2 min.
➢ Centrifugation of the samples at 12,000 × g for 10 min at 4°C is done.
➢ Following centrifugation, the mixture is separated into lower red, an interphase, and a colorless upper aqueous phase. RNA remains exclusively in the aqueous phase. The upper aqueous phase is carefully transferred into a fresh 1.5 mL tube without disturbing the interphase. Reextraction is carried out with an equal volume of chloroform:isoamyl alcohol and centrifuged at 10,000 g for 10 min at 4°C.
➢ The upper aqueous phase is carefully transferred into a fresh 1.5 mL tube without disturbing the interphase.
➢ The aqueous phase is processed for RNA precipitation.

Basic Life Science Methods. https://doi.org/10.1016/B978-0-443-19174-9.00028-3

➤ 1/3 volume of 8M LiCl is added to the aqueous phase, mixed by inversion, and stored at 20°C at least for 2 h and for better yield overnight.
➤ Centrifugation is done at 15,000 × g for 20 min.
➤ The supernatant is removed from the tube leaving only the RNA pellet.
➤ The pellet is washed with 1 mL of 75% DEPC ethanol.
➤ The samples are vortexed briefly and then the tubes are centrifuged at 7500 × g for 5 min at 4°C and the wash is discarded.
➤ The RNA pellet is air dried for 5—10 min.
➤ The RNA pellet is resuspended in RNase-free water (DEPC water).

After dissolving, samples are run in 1% agarose gel to check the quality of RNA (Fig. 28.1) and the presence of any DNA contamination, if present is removed by the method given below:

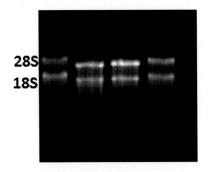

FIGURE 28.1

Representative picture of RNA showing 28s and 18s rRNA.

DNase treatment

➤ RNA sample is added with 5 μL of 10× buffer.
➤ 1 μL of DNase-1 is added.
➤ The sample is incubated at 37°C for 30 min.
➤ Finally, either 1 μL of EDTA (stop solution) is added or mixture is incubated at 65°C for 10 min.

The concentration of RNA and its integrity is verified by an optical density measurement ratio of OD260/OD280 and also by nanodrop.

cDNA synthesis

First-strand cDNA is synthesized from isolated RNA using oligo-dT and superscript II (Invitrogen) in a reaction volume of total 20 μL. cDNA synthesis is done using First-Strand cDNA Synthesis Kit using oligodT primers with some modifications.

The first-strand cDNA is synthesized by using manufacturer's protocol as follows. The cDNA synthesized is stored at $-20°C$. The protocol for synthesis of first strand is given below

- The following reagents are added in an RNase-free tube on ice as per the indicated order.

Template RNA	$1-2 \mu L$
Oligo(dT)$_{18}$ primer	$1 \mu L$
5× buffer	$4 \mu L$
Nuclease-free water	$7-8 \mu L$

- The components were mixed gently and a brief spin is given.
- Incubation at $65°C$ for 5 min is given.
- The tubes are chilled on ice and brief spin is given.
- The tubes are again placed on ice and the following components are added as per the indicated order.

RiboLock RNase inhibitor (20 U/μL)	$1 \mu L$
10 mM dNTP mix	$2 \mu L$
RevertAid M-MuLV RT (200 U/μL)	$2 \mu L$

- The components are mixed gently and a brief spin are given.
- The components are incubated at $42°C$ for 40 min.
- The reaction is terminated by heating at $70°C$ for 5 min.

Reference

Asif, M., Trivedi, P., Solomos, T., Tucker, M., 2006. Isolation of high-quality RNA from apple (Malus domestica) fruit. Journal of Agricultural and Food Chemistry 54 (15), 5227–5229.

DNA isolation from the bacterial culture broth

29

Chapter outline

Principle: The basic procedure for the DNA extraction from a cellular lysate relies upon the same principles regardless of the type of DNA being extracted (plasmid or chromosomal) or the type of organism used. A lysate is composed of all internal contents of the cell including the cytoplasm, all of the solubilized proteins and nucleic acids as well as all of the insoluble cell contents including cell wall, cell membrane, ribosomes, and other cellular organelles. There is one major difference between the extraction of DNA from the bacteria and fungi and plant cells and the extraction of DNA from animal cells. The bacteria, fungi, and plants, all have cell walls that first must be cracked open before cell lysis can occur. Animal cells however are surrounded by a plasma membrane that is much easier to disrupt without disturbing the cell cytoplasmic contents. In addition, prokaryotic cells are different from eukaryotic cells due to the absence of the nucleus and other membrane bound organelles in prokaryotes. The first step in the isolation of DNA from bacteria involves the cracking open of the cell wall. Fortunately, this is relatively simple because almost all eubacteria have a cell wall made of peptidoglycan. Peptidoglycan is a polymeric carbohydrate made up of two different sugar residues (N-acetyl-glucosamine and N-acetyl-muramic acid) which are repeatedly to form long chains. Peptidoglycan is degraded by an enzyme called lysozyme which catalyzes the splitting of the covalent bond between the N-acetyl glucosamine and N-acetyl muramic acid sugar residues causing the cell wall to become perforated. Once this is accomplished, the addition of a strong ionic detergent such as SDS, which is dissolved in a very basic solution of NaOH, causes the final dissolution of the cell wall and solubilizes to as a

Basic Life Science Methods. https://doi.org/10.1016/B978-0-443-19174-9.00029-5

clear lysate. The cleared lysate at this point has a very basic pH, contains a strong detergent, and has all of its large macromolecular substances in solution. The next step involves the extraction of the DNA from the rest of these cytoplasmic and cellular contents. In the extraction of plasmid we capitalize upon the size difference between the plasmid DNA and the chromosomal DNA to selectively precipitate the large components (cell wall, cell membrane, proteins, and ribosomes) away from the plasmid DNA and small RNA molecules. In order to facilitate this process, we add a solution containing a 3M potassium acetate salt solution suspended in acetic acid. The salt absorbs the water molecules disrupting the spheres of hydration, which is responsible for the solubilization of the macromolecular cellular components, and the acid in this solution reacts with the basic NaOH and neutralizes this solution. A white precipitate forms upon addition of the acidic salt solution that can be separated from the liquid phase containing the plasmid by centrifugation allowing the transfer of the clarified solution to another container. This clarified solution now contains the plasmid.

DNA, small-solubilized proteins and some small cellular RNAs (tRNA and rRNA). These smaller molecules can be recovered by the addition of 95% ethanol to a final concentration of 70%. At 70% concentration of ethanol and 3 M salt concentration plasmid DNA readily precipitates out of solution. Remember the precipitation reaction involves the drawing of the hydration spheres of water away from the DNA molecule causing the plasmid DNA to precipitate out of solution. The salt and ethanol in the precipitation reaction readily absorb the water away from the DNA for 1 min to reconstitute the plasmid DNA.

Materials and equipment (materials and reagents)

- 10 mM Tris-HCl buffer (pH-8.0) [act as buffer to stabilize the pH]
- EDTA buffer [act as chelating agent]
- Adjust pH to 8.0 by adding NaOH. EDTA does not dissolves until pH of solution is 8.0, so first adjust pH up to 8.0, then it dissolves
- TE buffer solution (pH 8.0) [long-term storage of nucleic acid]
- All equal volume of both the solution given below 10 mL—10 mM Tris-HCl, pH 8.0
- 10 mL—1 mM EDTA, pH 8.0
- SDS 10% solution [denature the protein and break its native confirmation]
- Phenol—chloroform isoamyl alcohol (PCI) solution (25:24:1) [separation of nucleic acid, protein, and phenolic compounds]
- Phenol—25 mL, chloroform—24 mL, and isoamyl alcohol—1 mL.

Step-by-step method details (experimental procedure)
Procedure

Pellet the bacterial from any one of the following methods:

1. If liquid culture is present, then take 1—2 mL of culture in centrifuge tube and if you have to harvest bacteria from agar plates, then scrap the bacterial growth with the help of loop and suspend it in 1—2 mL of TE buffer.
2. Centrifuge the tubes at 8000 rpm for 5 min.
3. Resuspend the pellet in 900 μL of TE buffer.
4. Add 1/10 volume of 10% SDS solution (100 μL) and mix by vortex.
5. Warm the tube to 50—60°C and incubate for approximately 2 h.
6. After 2 h the solution becomes thicker than it was in starting, now mix 600 μL of phenol:chloroform:isoamyl alcohol mixture (25:24:1) to it and shake gently by repeatedly inverting the tube. White precipitate appears when this mixture is added to the tube and after shaking solution becomes creamy orange (white due to protein precipitate and orange due to phenol).
7. Keep for 5 min and then centrifuge at 10,000 rpm for 10 min.
8. Upon centrifugation, three layers appears in tube. Upper aqueous layer is transparent like water, which contains DNA, below this is a thin layer of white color that is protein precipitate and the lower dark orange layer is phenol chloroform mixture.
9. Collect the upper transparent aqueous layer in fresh new eppendorf tube and discard the lower two layers. Keep the eppendorf tube in refrigerator to cool it for 10—15 min.
10. Now add double volume of chilled propanol to the aqueous DNA solution drop by drop from its wall. After adding, keep it again at 4°C for overnight so that DNA is precipitated easily.
11. After precipitation centrifuge the tube at 10,000 rpm for 10 min and discard the supernatant, before discarding mark the DNA pellet with marker on wall of tube.
12. Dry the DNA pellet in air or by keeping the tube in inverted position on filter paper.
13. Dissolve the dry DNA pellet in 50 μL 1×TE buffer.
14. Check the quality of the DNA isolated by agarose gel electrophoresis.

Expected outcomes

The DNA was isolated from culture broth inoculated with the bacterial species. The isolated procedure involves several steps. The DNA was obtained via cell lysis, removal of protein and lipids and precipitation as the main steps. The isolated DNA was kept in 50 μL of 1×TE buffer to prevent its degradation and was further analyzed via agarose gel electrophoresis.

Advantages

One of the most trusted, well-known and widely accepted methods of DNA extraction is our PCI method. We get good DNA purity and yields. This method is cheap, easy to use, and reliable.

Limitations

The phenol—chloroform method of DNA extraction is time-consuming and tedious. However, by standardizing it properly, we can use it routinely. Also, the process of chemical preparation is time-consuming and tedious too. The chemicals used in phenol—chloroform DNA extraction are dangerous for health which is the major limitation of the PCI method. The phenol is volatile and may cause skin burn and irritation. The chloroform makes you unconscious. Moreover, what you get depends on how you work, meaning if you don't have a good practical hand, you can't isolate good DNA.

Safety considerations and standards

All workers shall wear long sleeved lab coat, covered shoes, safety glasses/goggles, and nitrile gloves (or glove material impermeable and resistant to the substance). Wear appropriate face shield and insulated gloves if necessary. Know where the nearest emergency eyewash station and safety shower are in the laboratory before beginning work. Each equipment is routinely checked. (1) Ensure that a spill kit is well stocked and readily available should a spill occur. (2) Make sure you are using appropriate tubes for the procedure—polypropylene is best, and the cap must be very tight-fitting. Do not use polycarbonate tubes—these will be dissolved by the phenol—chloroform. Glass is not recommended due to risk of breakage. (3) Wash hands thoroughly with soap and water after work is completed. (4) Refer to the SOPs on the safe use of equipment: magnetic stirring plate, autoclave, biosafety cabinet, fume hood, centrifuges, water bath, etc. (5) Refer to the SDS prior to the use of chemicals.

DNA isolation from fungal mycelium

Chapter outline

Introduction

The handful of studies on microbial communities and fungal diversity is steadily increasing. Large-scale community studies frequently demand the identification of a large number of fungal species and strains, and genomic DNA must be obtained in a high-throughput way for the molecular identification of such a large number of fungal isolates. Fungal cultures produced in liquid broth or culture plates can have their genomic DNA extracted. The fungal mass from the culture plate must be scraped away using a fine spatula, and the fungal mass from the culture broth may be retrieved by filtering the broth through a 10 mL syringe containing glass wool, which allows the broth to flow through while preserving the fungal mycelia.

Procedure or steps

1. Take 20–30 mg of harvested Mycelia, blot dry between the paper layers and immediately froze it in liquid nitrogen.
2. The grounded mycelium should be resuspended and lysed in 500 mL of lysis buffer (40 mmol/L Tris-acetate, 20 mmol/L sodium acetate, 1 mmol/L EDTA, 1% w/v SDS pH7·8) or alternatively use CTAB buffer containing 200 mM Tris HCL pH 8.5, 250 mM NaCl, 25 mM EDTA, and 2% CTAB. Crush the mycelium in extraction buffer using pestle-mortar to make a slurry.
3. Transfer the mixture to a sterilized 2 mL microfuge tube and incubate at 65°C in a water bath for 1 h and vortex after every 10 min interval.
4. After incubation, centrifuge the lysate at 14,000 rpm for 10 min.
5. Transfer the Supernatant into 1.5 mL microfuge tube.
6. Add 2–4 µL of RNase (20 mg/mL) to cell suspension and again incubate for 10 min at 65°C with intermittent mixing.

Basic Life Science Methods. https://doi.org/10.1016/B978-0-443-19174-9.00030-1

131

7. After the RNase A treatment, add equal volume of phenol: chloroform: Isoamyl alcohol (25:24:1) and mix well, followed by centrifugation at 13,000 rpm for 10 min (Note: this step can be repeated once more to completely get rid of proteins/cell debris).

8. Add 30 μL of 3M sodium acetate, pH 5.2 and incubated at −20°C for 10−20 min.

9. Centrifuge the lysate at 4000 rpm at 4°C for 15 min

10. Transfer the supernatant to a clean microfuge tube.

11. Precipitate DNA by the adding one volume of isopropanol

12. Incubate for 15 min, pellet at 14,000 rpm for 10 min, and wash with ice cold 70% ethanol to remove salt contaminants

13. Air dry for 15−20 min and resuspend in 60 μL of Tris-EDTA buffer, pH 8.0 or nuclease-free water

14. Quantitatively and qualitatively check of DNA can be done using a nanodrop spectrophotometer and can be further diluted to a working concentration of 30 ng/μL and stored at −20°C for further use. DNA can be varified by runing on agarose gel (Fig. 30.1).

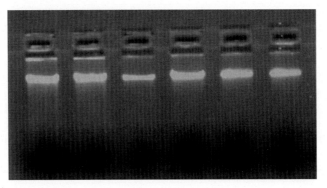

FIGURE 30.1

Representative pic of DNA isolated from fungi.

Precautions

- When doing the experiment, use safety glasses, a lab coat, and gloves. Ethanol and rubbing alcohol may cause blindness if ingested
- Protect your hands by wearing gloves. They should be changed often
- Use disposable equipment or thoroughly clean them before and after each sample
- Don't touch the place where you think DNA could be present
- Don't talk, sneeze, or cough in front of evidence, avoid touching your face, nose, and mouth.

Quantification of the DNA using spectrophotometer

31

Chapter outline

Principle: One of the most commonly used methods to determine the quantity of nucleic acids in a suspension depends upon the chemical characteristic of nucleic acids which allow them to absorb ultraviolet (UV) light strongly in the wavelength between 254 and 260 nm. Nucleic acids absorb in this range because of the nitrogenous base, pyrimidine and purine ring structures. Because nucleic acid absorbs UV light proportional to its concentration, it is easily quantified using this method. Previously, it was calculated by the use of standard concentrations of nucleic acids that 50 µg of DNA will give an absorption value of 1.0 at 260 nm and that 40 µg of RNA will give an absorption value of 1.0 at 260 nm. The failing of this quantitative method is the possible contamination of the DNA precipitation with RNA and/or with organic solvents containing ring structures such as phenol and chloroform. For this reason, critical DNA quantification should be performed upon highly purified DNA preparations only. However, by using UV absorption spectroscopy, it is possible to determine if a DNA preparation has protein contamination or not. When DNA quantification is performed by measuring the absorption of a solution at 260 nm, one also measures the absorption of the solution at 280 nm. The rationale for this procedure stems from the ability of protein to absorb UV light in the 260–280 nm range. Proteins absorb in this range due to the presence of the ring structures on the amino acid residues phenylalanine, tryptophan, and tyrosine. However, pure nucleic acid absorbs two times more strongly at 260 nm than at 280 nm.

Therefore, if you measure your nucleic acid at both 260 and 280 nm, pure DNA should have a 260/280 ratio of 2:1. If this ratio is smaller, DNA has some contaminating substances and should be quantified by some other method for accuracy.

Materials and equipment (materials and reagents)

- DNA sample
- Spectrophotometer
- Distilled water
- Cuvettes
- Micropipettes
- Microtips
- Tissue paper

Step-by-step method details (experimental procedure)
Procedure

1. Dilute plasmid extract from plasmid DNA by adding 10 μL of extracted DNA to 990 μL water in a glass test/cuvette tube and mix well.
2. Place 1 mL of your diluted plasmid in one quartz cuvette and 1 mL of water in the other.
3. Test the absorbance by placing blank cuvettes and sample in the UV spectrophotometer.
4. Note the reading at both the ultraviolet wavelength of 260 and 280 nm.
5. In order for your nucleic acid determination to be accurate, you must measure the ratio of the optical density (OD) at 260 versus 280 nm.
6. If DNA is present, this will add to OD 260 and subsequent quantification will be inaccurate since this method does not discriminate between RNA and DNA.

Expected outcomes
DNA quantification

The number that you read from the spectrophotometer is the optical density or OD. Use the following equation to quantify the amount of DNA or RNA in this mixture. OD 260 1.0 = OD 260 reading.

50 μg DNA × O.D.

Optical density of the diluted DNA sample was taken at 260 nm. The optical density was multiplied with 50 μg to get the quantity of DNA. After that, the OD was taken at 280 nm to check the purity of DNA sample by calculating the ratio of the 2 ODs.

The sample solution had an OD 260 = 0.2 OD 260 1.0 = 0.2.
X = 50 µg × 0.2 = **10 µg of DNA**.
# Purity of DNA calculated: **1.61**.

Advantages

The main advantage is it provides a simple and accurate estimation of the concentration of nucleic acids in a sample. Purines and pyrimidines in nucleic acid show absorption maxima around 260 nm if the DNA sample is pure without significant contamination from proteins or organic solvents. The ratio of OD260/OD280 should be determined to assess the purity of the sample.

Limitations

- It is a time-consuming process
- Low specificity
- Sensitive to contaminants

Safety considerations and standards

It is important that the DNA sample to be measured using this approach should be pure in order to avoid other interfering biological entities and proteins.

Formaldehyde gel electrophoresis of the isolated RNA sample

32

Chapter outline

Principle: Formaldehyde-agarose gel electrophoresis gives enhanced sensitivity for gel and subsequent analysis (e.g., northern blotting). A key feature is the concentrated RNA loading buffer that allows a large volume of RNA sample to be loaded onto the gel than conventional protocol.

Materials and equipment (materials and reagents)

- $1 \times$ TAE buffer
- Agarose gel (1.3%)
- Formaldehyde (37%)
- Ethidium bromide (1 mg/mL)
- Master mix for sample preparation
- Formaldehyde reacts with RNA, and when RNA is heated in the presence of formaldehyde, all secondary structures are removed. Hence, gel electrophoresis of RNA in agarose gels containing formaldehyde provides a good denaturing gel system.

Step-by-step method details (experimental procedure)
Procedure
1. Prepare 0.8% agarose gel in 22.55 mL of 1× TAE buffer and heat it up to 55–60°C for 30 min till it becomes transparent.
2. Add 2.45 mL of 37% (v/v) formaldehyde and swirl.
3. Add 1 μL of EtBr as fluorescent dye.
4. Pour the gel and allow agarose gel mixture to solidify for 30 min.
5. Add 5 μL of gel loading dye with 6 μL of sample.
6. By pipette mixing load RNA, sample into the wells.
7. Run the gel until bromophenol blue dye migrates almost two-third of the gel.
8. Observe the gel under Gel DOC System (Fig. 32.1).

Percent (%) formamide in the sample:

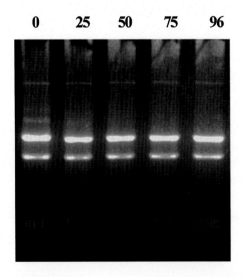

FIGURE 32.1

Formaldehyde gel electrophoresis.

Expected outcomes

The RNA isolated can be analyzed via formaldehyde gel electrophoresis. It can be prepared by using 0.8% agarose gel in 22.55 mL of 1× TAE buffer and 2.45 mL of formaldehyde along with ethidium bromide (μL).

Advantages

The key advantages include shortened run times, a fivefold reduction in formaldehyde concentration, a significantly improved resolution of long RNAs, and consistency in separation. Polyacrylamide gel electrophoresis (PAGE) is used for separating proteins ranging in size from 5 to 2000 kDa due to the uniform pore size provided by the polyacrylamide gel.

Limitations

It is a time-consuming process. The automated process becomes complicated and therefore costly. It has a low separation power.

Safety considerations and standards

Formaldehyde is corrosive and may cause serious burns; irritant contact to the skin will cause inflammatory effect; will cause one to become sensitized; classified as possible cancer-causing agent. Pore size is controlled by modulating the concentrations of acrylamide and bis-acrylamide powder used in creating a gel. Care must be used when creating this type of gel, as acrylamide is a potent neurotoxin in its liquid and powdered forms.

Paper chromatography of different amino acid

33

Chapter outline

Principle: Chromatography is an important technique which helps to resolve the closely related substances based on physical and chemical characteristics. The mixture of compounds is separated on the basis on partition coefficients. In this technique the stationary phase may be liquid or solid, whereas mobile phase may be liquid or solid or gas. Following section describes the method used by using paper chromatography for separation, quantification, and identification of analytes such as organic acids, sugars, amino acid, and plant-based metabolites. The analyte separation is based on the liquid–liquid partitioning operated on paper chromatographic setup. Other chromatographic alternative techniques used for separation include Liquid Chromatography (LC), Gas Chromatography (GC), Thin-Layer Chromatography (TLC), HPLC, etc.

Materials required for setup of technique

- Whatmann No. 1 filter paper
- Chromatography chamber
- Hair-dryer or spot-lamp
- Atomizer
- Microsyringe or micropipette
- Mobile phase (solvent system)

Basic Life Science Methods. https://doi.org/10.1016/B978-0-443-19174-9.00033-7

Experimental procedure

1. In the first step the chromatographic sheet is cut carefully into convenient size (40 × 24 cm).
2. A straight line is drawn by using carbon pencil approximately 5 cm away from the base of the rectangular sheet and then mark the spots in the line at intervals of about 1 cm (as shown in figure below).
3. On these spots, now apply small volume of analytes by using capillary tubes or micropipettes.
4. The drying of spots is accomplished by using hair-dryer. This step is done for obtaining better resolution and separation of analytes in the applied mixture.
5. The sample loaded sheet is now placed in stainless steel trough hanged in a chromatographic chamber. Make sure that the sample spotted end of the sheet should descend in the trough (refer figure). The paper must be kept in such a way that it is immersed in a solvent placed in the bottom of chamber or in a Petri dish.
6. The chamber was closed by placing a lid to avoid the evaporation of the solvent.
7. The setup was stopped once solvent reaches to other end of the paper. This step is critical for developing a highly resolved chromatogram.
8. If the analyte is amino acid mixture, one can spray ninhydrin reagent on chromatogram by using atomizer.
9. The chromatogram was dried at room temperature for 5 min followed by incubation for 2–3 min at 100°C in oven.
10. After step 9, you will observe purple spots representing the amino acid, if yellow spots are observed the representative amino acid in that spot are proline and hydroxy proline.
11. The rf (retention factor) of each spot was calculated by using following formulae

$$\frac{\text{Distance(cm)moved by the solute from the origin}}{\text{Distance(cm)moved by the solvent from the origin}}$$

12. The identification of amino acid is done by calculating the rf values of each spot by comparing with standard spots developed parallelly in the chromatogram.
13. For quantitative estimation, each spot in the paper chromatogram was cut in small bits and immersed in the sterilized test tube.
14. To the above test tube, add 3 mL of elution mixture and shake vigorously for about 15 min.
15. The liquid is then decanted and the paper pieces are then eluted with 2 mL elution mixture. The step was repeated till paper gets colorless.
16. The eluate was cleared by centrifuging at 10,000 rpm for 10 min.
17. The intensity of the purple color was detected at 570 nm using colorimeter.

Biosafety considerations and laboratory standards

- Dry the spots paper chromatogram before pacing in chamber
- The spots where sample is loaded must not get immersed in the solvent
- The paper chromatogram must not touch the sides of the chamber
- The chamber must be closed to avoid the evaporation of the solvent
- The experimental setup must be operated in standard laboratory conditions for better resolution

The experimental outcome

The resolution of amino acids is quantified as follows by the application of 2% ninhydrin solution.

In addition, Rf value were calculated.

Rf value of Alanine = 5.6/6.6 = 0.85.

Rf value of Glutamine = 4.8/6.6 = 0.72.

Rf value of Leucine = 4.1/6.6 = 0.62.

Rf value of mixture of all the above = 5.2/6.6 = 0.79.

Advantages

- ✔ The technique is rapid
- ✔ Less amount of analyte is used
- ✔ The materials used in this technique are very cheap and affordable, hence can be operated in laboratories at elementary level
- ✔ Very less operating space is required for carrying the experiment
- ✔ Resolving power of analytes is high.

Limitations

- ❖ Large samples are not resolved
- ❖ Quantitative analysis is poor compared to other techniques
- ❖ Very complex mixtures are not resolved efficiently
- ❖ Accuracy and efficiency are very less as compared to other techniques such as HPLC, UPLC, etc.

Separation of plant pigments using column chromatography

34

Chapter outline

Basic principle behind separation of biomolecules

The technique is employed to separate the biomolecules from a gross mixture of diverse compounds. The technique is operated for preparative separation of biomolecules ranging from micrograms to kilograms. The critical advantages of this technique lie in its low cost for operation and the disposability of stationary phase. Hence, making it sure that operations are simple and very less chances of cross contaminations.

Materials and equipment (materials and reagents)

- Micropipette
- Microtips
- Beaker
- Distilled water
- Acetone
- Methanol
- Silica gel
- Plants pigments
- Sample

Basic Life Science Methods. https://doi.org/10.1016/B978-0-443-19174-9.00034-9

Chemical solutions used for column chromatography

- Add silica gel (5 gm) (stationary phase)
- Acetone solution (stationary phase)
- Methanol solution (mobile phase)
- Adjust the volume to 1 L using distilled water.

Step-by-step procedure

1. Preparation of plants pigments (without acetone):
   1.1 The prepared slurry of silica gel was immersed into the sterilized glass or steel column.
   1.2 The assembly was left standby for 2–3 h such that slurry in column will settle for 2–3 h.
   1.3 The pigment sample was injected into the column for initiating the elution of pigments.
   1.4 The eluents were collected in separate tubes at the intervals of 2 min until whole pigment sample is eluted from the column.
2. Prepared samples of plant pigments (acetone):
   2.1 The prepared slurry of silica gel this time with acetone was immersed into the sterilized glass or steel column.
   2.2 The assembly was left standby for 2–3 h such that slurry in column will settle for 2–3 h till acetone forms a layer above silica gel.
   2.3 The pigment sample was injected into the column for initiating the elution of pigments.
   2.4 Once loaded the sample starts diffusing into the gel and then keep adding methanol 1 mL on it as mobile phase.
   2.5 The above step resulted in formation of different layer of pigments.
   2.6 Fractions of solution (mobile phase) were collected at an interval of 2 min until whole pigment sample is eluted from the column.

Expected outcomes

The pigments were obtained at ease with differential colors such as, yellow, light green, green, and dark green in color.

Advantages of column chromatography

Different kinds of mixtures can be separated by column chromatography including large quantities of samples. The technique is robust and the separated analytes can be reused. Moreover, the process of separation is completely automated.

Main limitations

- Process is time-consuming.
- For higher quantities of solvents, the process is expensive.
- In certain cases, automated setup gets complicated.
- The separation power is low.

Safety concerns and experimental standards

(1) The drop applied to the paper must be allowed to dry before the paper is placed in the solvent to be run; (2) the drop must not be so near the PAGE 103 bottom of the paper that it is immersed in the solvent; (3) the paper must not touch the sides of the test-tube except at its four corners, (4) the tube must be corked so that the atmosphere about the paper is saturated with the solvents used.

Alternative methods/procedures

Other techniques for Liquid Chromatography, Gas Chromatography, Thin-Layer Chromatography, HPLC, etc.

Separation of plant pigment by thin-layer chromatography

35

Chapter outline

Theory: Thin-layer chromatography (TLC) is a chromatography technique used to separate mixtures. TLC is performed on a sheet of glass, plastic, or aluminum foil, which is coated with a thin layer of adsorbent material, usually silica gel, aluminum oxide, or cellulose (blotter paper). This layer of adsorbent is known as the stationary phase. After the sample has been applied on the plate, a solvent or solvent mixture (known as the mobile phase) is drawn up the plate via capillary action. Because different analytes ascend the TLC plate at different rates, separation is achieved. TLC can be used to monitor the progress of a reaction, identify compounds present in a given mixture, and determine the purity of a substance.

Materials required

- TLC plate
- TLC chamber
- Capillary tubes
- Reagent spray bottle
- Conical flasks
- Beakers

Basic Life Science Methods. https://doi.org/10.1016/B978-0-443-19174-9.00035-0

Reagents

✔ 2 mL of plant pigment in different tubes
✔ Solvent mixture of hexane and ethyl acetate in the ratio 7:3 by volume

Procedure

1. Pour the solvent mixture into the TLC chamber and close the chamber.
2. The chamber should not be disturbed for about 30 min so that the atmosphere in the jar becomes saturated with the solvent.
3. Cut the plate to the correct size and using a pencil (never ever use a pen) gently draw a straight line across the plate approximately 1 cm from the bottom.
4. Using a micropipette, a minute drop of plant pigments is spotted on the line.
5. Allow the spot to dry.
6. Spot the second plant pigments on the plate [enough space should be provided between the spots].
7. Repeat the above step for spotting the unknown plant pigments.
8. Place the plate in the TLC chamber as evenly as possible and lean it against the side (immerse the plate such that the line is above the solvent). Allow capillary action to draw the solvent up the plate until it is approximately 1 cm from the end.
9. Remove the plate and immediately draw a pencil line across the solvent top.
10. Under a hood, dry the plate with the aid of a blow dryer.
11. Spray the dry plate with ninhydrin reagent.
12. Dry the plates in hot air oven at 105°C for 5 min.
13. After some time, mark the center of the spots, then measure the distance of the center of the spots from the origin and calculate the Fro values.

R_f value can be calculated using the formula:

R_f = distance moved by the substance from origin/distance moved by solvent from origin.

Observation and results

Silica gel along with distilled water was used as stationary phase, poured on TLC plate. Plant leaf pigments were used as sample and hexane and ethyl acetate were used as mobile phase. Sample were loaded on TLC plate and TLC plate was dipped in mobile phase. Mobile phase ran on the stationary phase with sample pigments. Different pigments get separated on the TLC plate. Separated pigments where yellow, light yellow, orange, light orange, green, light green, and dark green (Fig. 35.1).

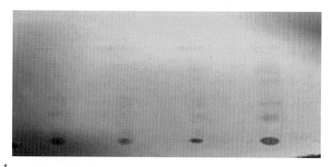

FIGURE 35.1

Different pigments separated on TLC plate.

Chromatin immunoprecipitation protocol

36

Chapter outline

Proteins, DNA, and RNA constitute chromatin. It is found in the nucleus of eukaryotic cells and regulates cell-specific or tissue-specific gene expression as well as DNA replication and repair, among other things. The ChIP test has evolved into one of the most practical and valuable methods for studying gene expression, histone modification, and transcription regulatory processes. Furthermore, ChIP tests are particularly valuable for identifying transcription factors and the genes that they target. This test examines if a certain protein—DNA interaction exists at a specific place, condition, and time. The most important aspect in a successful ChIP experiment is the selection of a suitable antibody for the immunoprecipitation stage.

When processing up to 5—10 distinct samples, the ChIP methodology given here may be completed in roughly 2.5 h. It avoids the time-consuming crosslinking-reversal, overnight antibody incubation, and antibody capture with protein A/G agarose employed in typical procedures. In an ultrasonic bath, antibodies are incubated. The biotin-streptavidin technique is used to capture antibodies. This procedure is comparable to previous fast/sensitive ChIP methods that have been described. This approach uses a chelating resin solution and silica-based columns to clean up DNA samples. These methods drastically minimize the amount of time it takes to complete the experiment. List of chemicals/reagents required are given in table.

S. No.	Chemicals/reagents	S. No	Chemicals/reagents
	Biotinylated antibody or non–biotinylated antibody plus anti-IgG-biotin		Lysis buffer, dilution buffer, wash buffer, and chelating resin
	Biotinylated normal IgG (negative control)		37% formaldehyde solution
	Leupeptin		1M glycine
	Aprotinin		PCR kit
	Phenylmethylsulphonyl fluoride		Phosphate-buffered saline
	Streptavidin magnetic beads or agarose beads		Dimethyl sulfoxide
	DNA purification kit		Deionized or distilled water
	Optional: primers for a known target gene		

Materials

Pipettes and pipette tips
15 mL falcon tubes
Ultrasonic bath
Eppendorf tube rotator
Refrigerated ultracentrifuge
Water bath and heatblock

1.5 mL microcentrifuge tubes
Sonicator
Rocking or shaking device
Benchtop ultracentrifuge
Benchtop centrifuge
PCR thermocycler

Procedure

1. Incubate the cells for 15 min at room temperature with 37% formaldehyde diluted to a 1% final concentration on a rocking or shaking device.
2. Add 1M glycine diluted to a final concentration of 125 mM to quench the formaldehyde. Pellet the cells and remove the medium after 5 min of rocking at room temperature (at this point, samples can be kept overnight at $-70\,°C$).
3. 10 g/mL Leupeptin, 10 g/mL Aprotinin, and 1 mM PMSF are added to the lysis buffer as protease inhibitors. For every 5×10^6 cells, resuspend the cell pellet in 500 mL lysis buffer. Pipette the cells up and down to resuspend them, then cool them for 10 min (keep samples on ice from this step forward).
4. Sonicate the samples to shear the chromatin to a length of roughly 1 kb on average. 500 μL of each sample should be transferred to a 1.5 mL microcentrifuge tube.
5. Centrifuge the lysates for 10 min at 12,000 × g in a refrigerated ultracentrifuge. Remove the pellet and collect the supernatant in a clean tube.

6. Add 1 mL of dilution buffer (containing the same proportion of protease inhibitors as in step 3) to the supernatant, then 5 ug of antibody or normal IgG to the samples. Incubate in an ultrasonic bath at room temperature for 15 min (or overnight at 2−8°C on a rotating device if the antigen is predicted to have a low degree of expression). Add 5 ug of secondary antibody and incubate for 15 min in an ultrasonic bath at room temperature (alternatively, incubate for 1−2 h on a rotating device at 2−8°C). When utilizing a biotinylated primary antibody, the secondary antibody process is not required.

7. In a rotating device, add 50 μL of Streptavidin beads (magnetic or agarose) to the samples and rotate for 30 min at 2−8 °C.

8. If you're using magnetic beads, place the tube in the magnet for 2 min to gather the beads. If you're using agarose beads, gather them by centrifuging them for 1 min at 12,000 × g. 4 washes using Wash Buffers that have been pre-chilled to 2−8°C Add 1 mL of Wash Buffer to each wash. Begin with Wash Buffer 1 and work your way up to Wash Buffer 4. Between each wash, pipette the beads up and down.

9. Add 100 μL of chelating resin solution carefully to the beads after the previous wash and pipette up and down for around 10 s. Using a heatblock or a temperature-controlled water bath, boil the sample for 10 min.

10. Centrifuge at 12,000 × g for 1 min at room temperature and transfer the supernatant (~80 μL) to a clean microcentrifuge tube.

11. Refill the beads with 120ul of distilled or deionized water. Pipette up and down for 10 s, centrifuge for 1 min, collect the fresh supernatant, and combine it with the supernatant from Step 10 (samples can be kept at −20°C or −70°C at this stage).

12. Using a DNA purification kit, clean up and concentrate the DNA preparation. In 50 μL of deionized or distilled water, resuspend the DNA. This step boosts the yield of PCR fragments by concentrating the DNA in a smaller volume and reducing contaminants that might interfere with the PCR reaction (samples can also be kept at −20 °C or −70 °C at this stage).

13. 2−10 μL of the DNA sample can be used in the PCR reactions.

Precautions

(a) The Dilution and Wash Buffers should be maintained at 2−8°C. It's critical to keep the Dilution and Wash Buffers cold before using them.

(b) Formaldehyde is very poisonous, combustible, and potentially carcinogenic.

(c) The Lysis Buffer should be kept at room temperature.

Further reading

Chromatin Protocols. In: Becker, P.B. (Ed.), 1999. Methods in Molecular Biology, Volume 119. Springer.

Becker, P.B. (Ed.), 1999. Chromatin Protocols. Methods in Molecular Biology, Volume 119. Springer.

Bernstein, B.E., Kamal, M., Lindblad-Toh, K., Bekiranov, S., Bailey, D.K., Huebert, D.J., Lander, E.S., 2005. Genomic maps and comparative analysis of histone modifications in human and mouse. Cell 120 (2), 169—181.

Cosma, M.P., Tanaka, T., Nasmyth, K., 1999. Ordered recruitment of transcription and chromatin remodeling factors to a cell cycle—and developmentally regulated promoter. Cell 97 (3), 299—311.

Dahl, J.A., Collas, P., 2008. A rapid micro chromatin immunoprecipitation assay (ChIP). Nature Protocols 3 (6), 1032—1045.

Dundr, M., Hoffmann-Rohrer, U., Hu, Q., Grummt, I., Rothblum, L.I., Phair, R.D., Misteli, T., 2002. A kinetic framework for a mammalian RNA polymerase in vivo. Science 298 (5598), 1623—1626.

Fesenfeld, G., Groudine, M., 2003. Controlling the double helix. Nature 421 (6921), 448—453.

Johnson, K.D., Bresnick, E.H., 2002. Dissecting long-range transcriptional mechanisms by chromatin immunoprecipitation. Methods 26 (1), 27.

Kuo, M.H., Allis, C.D., 1999. In vivo cross-linking and immunoprecipitation for studying dynamic protein: DNA associations in a chromatin environment. Methods 19 (3), 425—433.

Li, C.C., Ramirez-Carrozzi, V.R., Smale, S.T., 2006. Pursuing gene regulation'logic'via RNA interference and chromatin immunoprecipitation. Nature Immunology 7 (7), 692—697.

Medeiros, R.B., Papenfuss, K.J., Hoium, B., Coley, K., Jadrich, J., Goh, S.K., Ni, H.T., 2009. Novel sequential ChIP and simplified basic ChIP protocols for promoter co-occupancy and target gene identification in human embryonic stem cells. BMC Biotechnology 9 (1), 1—14.

Nandiwada, S.L., Li, W., Zhang, R., Mueller, D.L., 2006. p300/Cyclic AMP-responsive element binding-binding protein mediates transcriptional coactivation by the CD28 T cell costimulatory receptor. The Journal of Immunology. 177 (1), 401—413.

Nelson, J.D., Denisenko, O., Sova, P., Bomsztyk, K., 2006a. Fast chromatin immunoprecipitation assay. Nucleic Acids Research 34 (1), e2.

Nelson, J.D., Denisenko, O., Bomsztyk, K., 2006b. Protocol for the fast chromatin immunoprecipitation (ChIP) method. Nature Protocols 1 (1), 179—185.

Roh, T.Y., Ngau, W.C., Cui, K., Landsman, D., Zhao, K., 2004. High-resolution genome-wide mapping of histone modifications. Nature Biotechnology 22 (8), 1013—1016.

Solomon, M.J., Varshavsky, A., 1985. Formaldehyde-mediated DNA-protein crosslinking: a probe for in vivo chromatin structures. Proceedings of the National Academy of Sciences 82 (19), 6470—6474.

Types of media

37

Media: Any liquid or solid preparation is particularly designed for the growth, storage, or transportation of microbes or other types of cells. Differential media, selective media, test media, and specified media are examples of media that may be used to cultivate certain bacteria and cell types.

Microbes are grown in laboratories on a culture medium, which provides their nutritional needs. These needs change for various bacteria, hence scientists have devised a variety of culture medium formulas to acquire the necessary microbial strain.

Simple or basal media, such as Nutrient Broth, is made up of sodium chloride, peptone, meat extracts, and water.

Complex media comprises an extra unique element that aids in the enhancement of a certain trait or provides nutrients for the development of specific bacteria. It may include plant, animal, and yeast extracts such as blood, yeast extracts, serum, milk, meat extracts, soybean digests, and peptone.

Synthetic or defined media: It is utilized in research. They are made by following a precise recipe and combining distilled water with precise proportions of inorganic and organic ingredients.

Special media: The basic medium promotes the growth of a diverse range of microbiological forms. However, only a certain type or strain of bacteria may be cultured and isolated using a specified growth environment. Special media are prepared media used to cultivate a specific microbe. It is further classified into many categories.

(a) **Anaerobic media:** It contains substances that help anaerobic bacteria flourish. Robertson's fried beef media is one example.
(b) **Selective media** restricts the development of some germs while enabling others to thrive. Desoxycholate citrate medium for dysentery bacilli or mannitol salt agar with 7.5% NaCl for *Staphylococcus* are two examples.

(c) **Differential media:** This medium supports the development of several bacterial species and differentiates them based on their size, shape, color, or the generation of gas bubbles or precipitates in the medium. MacConkey medium and blood agar are two examples.

(d) **Enriched media:** It comprises complex organic molecules such as hemoglobin, serum, blood, or growth factors that help specific microorganisms proliferate. Blood agar (which is commonly used to grow certain *Streptococci* and other pathogens) and chocolate agar are two examples.

(e) **Transport media:** It is a buffer solution comprising peptone, carbohydrates, and other nutrients (excluding growth factors) to keep the bacteria alive throughout transit while preventing their multiplication. The Stuart medium for gonococci is one example.

(f) **Indicator media:** It incorporates a color indicator that changes when a certain bacterium grows on the medium. When *Salmonella typhi* colonies develop on the Wilson and Blair medium, the addition of sulfite changes the color to black.

Preparation of nutrient agar media

Chapter outline

Principle

Nutrient agar is a complex medium because it contains ingredients that contain unknown amounts or types of nutrients. Nutricomplex medium contains Beef Extract (0.3%), Peptone (0.5%), and Agar (1.5%) in water. The beef extract is the commercially prepared dehydrated form of autolyzed beef and is supplied in the form of beef paste. Peptone is casein (milk protein) that has been dautolyzedith with the enzyme pepsin. Peptone is dehydrated and supplied as a powder. Peptone and beef extract contains a mixture of amino acids and peptides. Beef extract also contains water-soluble digest products containing other macromolecules (nucleic acids, fats, polysaccharides) as well as vitamins and trace minerals. There are many media ingredients that are complex: yeast extract, tryptone, and others. The advantage of complex media is that they support the growth of a wide range of microbes. Agar is purified from red algae which is an accessory polysaccharide (polygalacturonic acid) of their cell walls. Agar is added to microbiological media only as a solidification agent. Agar for most purposes has no nutrient value. Agar is an excellent solidification agent because it dissolves at near-boiling but solidifies at 45°C. Thus, one can prepare molten (liquid) agar at 45°C, mix cells with it, then allow it to solidify thereby trapping living cells.

Basic Life Science Methods. https://doi.org/10.1016/B978-0-443-19174-9.00048-9

Materials and equipment (materials and reagents)

- Electronic or beam balances
- pH paper or pH meter with standard buffers
- Graduated cylinder, 250 mL
- 250 mL Erlenmeyer flasks
- Spatula
- Nutrient agar
- Distilled water
- Petri dishes

Nutrient agar medium standard composition (1000 mL)

- **Beef Extract/Yeast Extract:** 1.5 gm (beef extract powder is a dehydrated extract of bovine tissue, which act as carbon source for growth of microbes)
- **Peptone:** 5 gm (a mixture of polypeptides and amino acids formed by the partial hydrolysis of protein, which act as nitrogen source for growth of microbes)
- **Agar:** 15 gm (solidifying agent)
- **NaCl:** 5 gm (mixing proportions $\sim$ cytoplasm of organisms)

Step-by-step method details (experimental procedure)
Procedure

1. Put the weighed amount of NA powder into 250 mL conical flask.
2. Add more distilled water to make 100 mL.
3. Adjust pH of the media to 7.0 using a pH meter by adding either acid or alkali as the case may be.
4. Autoclave at 121°C and 15 lbs pressure for 15 min.
5. Allow the flask to cool until the flask can be held by hand.
6. Pour the media into Petri dishes quickly under aseptic conditions.
7. Allow the media in plates to solidify, after solidification plates must be kept at room temperature or inside an incubator for future use.

Expected outcomes

25 mL of nutrient agar media was prepared using peptone, yeast extract, sodium chloride, agar, distilled water (Fig. 38.1). Peptone was used as nitrogen source for growth of microbes, yeast extract as a carbon source, NaCl to maintain the ionic concentration, and agar was used as solidifying agent. The media was autoclaved and used for further purposes.

FIGURE 38.1

Nutrient agar media.

Advantages

Nutrient agar is a general-purpose, nutrient medium used for the cultivation of microbes supporting the growth of a wide range of nonfastidious organisms. Nutrient agar is popular because it can grow a variety of types of bacteria and fungi, and contains many nutrients needed for bacterial growth.

Limitations

1. Individual organisms differ in their growth requirement and may show variable growth patterns in the medium.

2. Each lot of the medium has been tested for the organisms specified on the CoA. It is recommended for users to validate the medium for any specific microorganism other than mentioned in the CoA based on the user's unique requirement.

Safety considerations and standards

Wear protective gloves/protective clothing/eye protection/face protection. Follow good microbiological lab practices while handling specimens and culture. Standard precautions as per established guidelines should be followed while handling clinical specimens. Safety guidelines may be referred in individual safety data sheets.

Serial dilution of the bacterial cells

39

Chapter outline

Principle: Dilutions of bacterial cells and phage lysate are often necessary to obtain numbers of colonies or plagues that can be accurately counted. If too many cells are plated, a lawn of bacteria or one confluent plague is created on the surface of an agar plate. As you might think, this is useless for obtaining cell number counts or pure cultures from isolated colonies. One must learn this process. If the bacterial culture is diluted 1×10^{-5}, this means that the cells are diluted 1/100,000. To obtain this dilution, you could mix 1 mL of cells into 99,999 mL of buffer or media. However, we do not have any 100-L flasks. Additionally, if you are using expensive molecular biology grade reagents or growth mediums, this type of dilution would be prohibitively expensive. An alternative approach would be to perform a serial dilution.

Materials and equipment (materials and reagents)

- Test tubes
- Bacterial culture
- Cotton plugs
- Spirit lamp
- Distilled water

Basic Life Science Methods. https://doi.org/10.1016/B978-0-443-19174-9.00041-6

Step-by-step method details (experimental procedure)
Procedure

A serial dilution may be accomplished as follows:

1. Take six test tubes and label them as TT_1 TT_2 TT_3 TT_4 TT_5 TT_6.
2. Add 9 mL buffer in all six test tubes.
3. Make a 1/10 dilution to TT_1; this may be done by adding 1 mL of culture to 9 mL of buffer or other fluid and mixing the diluted culture well.
4. Now remove 1 mL from the 1/10 dilution and add it to TT_2 and mix well. This solution is now diluted 1/100 or 1×10^{-2}. Remove 1 mL from this 1/100 dilution, place it in TT_3, and mix well.
5. This tube is now diluted 1/1000 or 1×10^{-3}. Repeat this process two more times and you will have the desired dilution of 1×10^{-6}.
6. This procedure uses only 50 mL of diluents as opposed to 100,000 mL.
7. Example—1 mL in 9 mL = 1/100 or 1×10^{-2} and 1 mL of this dilution into 9 mL of diluents of TT_4 = 1/10,000 or 1×10^{-4}.

Advantages

1. Fewer errors.
2. Easier and faster preparation of calibration standards.
3. Calibrations solutions more evenly spaced.
4. Greater variability in calibration range.

Limitations

Serial dilution processes face two major challenges. The first is error propagation across columns or rows. With each sequential serial dilution step, transfer inaccuracies lead to less accurate and less precise dispensing. The result is that the highest dilutions will have the most inaccurate results.

Safety considerations and standards

1. Wear a lab coat and use eye protection.
2. Label the test tubes or bottles with the appropriate dilution, 10^{-1}–10^{-7}.
3. Make sure that the lid of the culture tube is firmly attached then shake the culture vigorously to separate clumps of cells and to distribute the organisms evenly throughout the liquid.

4. Remove a sterile syringe or pipette tip from its pack/container. Do not touch the parts which will come in contact with organism. If using a pipette tip, carefully attach to the dispenser.
5. Using aseptic technique (i.e., flaming the neck of the tube or bottle after removal and before replacement of its lid), remove exactly 1 cm^3 of the fluid and transfer to the dilution tube next in the series.
6. Place syringe or pipette tip into discard jar.
7. Mix the dilution well.
8. Repeat steps 4—7 until the last dilution tube is reached.

Spread plate method of the bacterial cells

40

Chapter outline

Principle: A way of separating bacteria cells on the agar surface is to obtain the colonies on the separate plate method. To inoculate a Petri dish, provide a simple and rapid method of diluting the sample by the mechanical method as the loop is spread over agar by a specialized glass rod, more and more bacteria are rubble off until individual bacteria is deposited on the agar. After inoculation, the area at the beginning of the spread pattern should show discrete colonies.

Materials and equipment (materials and reagents)

- Mixed culture in broth
- Inoculating loop
- Inoculating L-bar
- Nutrient agar media

Step-by-step method details (experimental procedure)
Procedure

1. Sterilize the loop using the appropriate aseptic technique; remove the loop full of broth from the mixed culture tube
2. Touch the loop full in the culture of the labeled Petri dish

3. Spread the culture using the sterile specialized glass rod called Labor prevents over-crowding of the microbes (the L bar is sterilized by dipping it in alcohol followed by 24–48 h)
4. Observe for growth and record your result

Expected outcomes

A particular concentration of the diluted sample (10^{-3}) was used for the spread plate technique. The sample was spread on the fresh media and the plate was incubated for 24 h at 37°C.

Advantages

1. In spread plate method, the isolation of the organism is easy because no subsurface colonies appear in this method.
2. Heat-sensitive bacteria are not affected.

Limitations

1. The optimum method for aerobes while microaerophilic tends to grow slower.
2. The dilutions should be accurate.
3. The volume of inoculum greater than 0.1 mL on the nutrient agar does not soak well and may result in colonies to coalesce as they form.
4. Enumeration difficulty can be formed if the colonies are crowded.

Safety considerations and standards

The protocol should be followed under all aseptic conditions preferably in Laminar Air Flow (safety cabinet) to avoid any contamination. Accurately measure the quantity while preparing the serial dilutions of the specimen. Allow the media plated to uniformly solidify. Do not inoculate the specimen on partially solidified media plates. Accurately measure the quantity of diluted specimen while inoculating onto the solidified media plates. Uniformly spread the specimen onto the media plate to get the discrete and well-developed colonies. Sterilize the L-shape Glass Spreader each time after using it to avoid any errors and contaminations (Fig. 40.1).

Pipette inoculum onto surface of plate

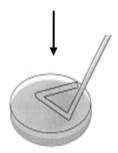

Spread evenly over the agar surface

Incubate

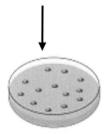

Colonies grow only on surface of media

FIGURE 40.1

Procedure for spread plate method.

CHAPTER

Streak plate method

41

Chapter outline

Principle: This streak plate method is for obtaining discrete colonies from a mixed culture or population of microbes. In this method, a sterilized loop or transfer needle is dipped into a suitable diluted suspension of organisms, which is then streaked on the surface of an already solidified agar plate to make a series of parallel, nonoverlapping streaks. The aim is to obtain colonies of microorganisms that are pure.

Materials and equipment (materials and reagents)

- Petri plates
- Inoculation loop
- Spirit lamp
- Nutrient agar
- Cling film/para film (used for sealing the plate to avoid microbial contamination from air)
- Bacterial colonies

Basic Life Science Methods. https://doi.org/10.1016/B978-0-443-19174-9.00040-4

Step-by-step method details (experimental procedure)
Procedure
1. Label all the plates, on the bottom, with the name of the organism to be inoculated by using a marker.
2. Sterilize the loop using the appropriate aseptic technique and wait till the palm is bearable.
3. Hold the tube containing the broth in the left hand and loop holding in the right hand.
4. Remove the cotton plug by using the little finger of the right hand and immediately flame the mouth of the tube.
5. Introduce the loop into the broth and withdraw one loopful of culture.
6. Flame the mouth of the tube, replace the cotton plug, and place the tube in the test tube rack.
7. Lift the Petri plate cover with the left hand and hold it at an angle of 60°C.
8. Streak the inoculum from side to side in parallel lines across the surface area.
9. Reflame, cool the loop, and turn the Petri plate to 90°C. Touch the loop to a corner of the culture in area 1 and streak the inoculum across the agar in area 2 as shown in figure.
10. The rest of the agar surface is now used to complete the streaking.
11. Incubate the plate at 35°C in an inverted position.

Expected outcomes
Pure culture isolation was obtained via transferring a loopful of particular colony from the spreading plate to the fresh media and incubated at 37°C for 24 h.

Advantages
1. For isolation of bacterial cultures into distinct separate colonies.

Limitations
1. There is a higher probability of contamination prior to isolation.
2. Streak plate method can be used for qualitative and not quantitative studies because it cannot be used for the enumeration of the approximate number of bacteria in the given sample.
3. The amount of inoculum added is not a measured quantity.
4. Colony count method cannot be applied on streak plate, since except in the fourth quadrant, isolated colonies are not formed.

Safety considerations and standards

The protocol should be followed under all aseptic conditions preferably in Laminar Air Flow (safety cabinet) to avoid any contamination. Accurately measure the quantity while preparing the serial dilutions of the specimen. Allow the media plated to uniformly solidify. Do not inoculate the specimen on partially solidified media plates. Accurately measure the quantity of diluted specimen while inoculating onto the solidified media plates. Uniformly spread the specimen onto the media plate to get the discrete and well-developed colonies. Sterilize the L-shape glass spreader each time after using it to avoid any errors and contaminations.

Identified bacterial colony in culture broth

42

Chapter outline

Principle: Preparing culture broth gives us the advantage of storing our bacteria for a longer duration. The nutrient broth has been considered a basic liquid media for the cultivation of bacteria.

Materials and equipment (materials and reagents)

- Peptone 5 g (a mixture of polypeptides and amino acids formed by the partial hydrolysis of protein, which acts as the nitrogen source for the growth of microbes)
- Beef extract: 1.5 g (Beef Extract Powder is a dehydrated extract of bovine tissue, which act as the carbon source for the growth of microbes)
- Distilled water 1000 mL (acts as solvent)
- NaCl acts gm
- Test tubes
- Bacterial culture plates

Step-by-step method details (experimental procedure)
Procedure

1. Put the weighed amount of peptone, NaCl, and beef extract in a test tube.
2. Heat with agitation to dissolve the constituents properly until the media looks almost transparent.

Basic Life Science Methods. https://doi.org/10.1016/B978-0-443-19174-9.00045-3

3. Add more distilled water to make 10 mL.
4. Adjust pH of the medium to 7.0, using a pH meter by adding either acid or alkali as the case may be.
5. Aliquot (divide into parts) as per your requirement.
6. Apply cotton plugs.
7. Autoclave at 121°C 15 lbs pressure for 15 min.
8. Allow the autoclave to cool.
9. Inoculate the test tube with culture from the streaking plate.
10. Incubate for 24 h at 37°C.
11. Store it at a cold temperature to inhibit its growth and for use in the future.

Expected outcomes

The turbidity obtained after the incubation time confirms the growth and presence of microbes in the inoculated culture broth.

Advantages

Broth cultures are convenient for growing a large number of bacteria very quickly. These bacteria for other tests combine them with freeze-medium for long-term storage, or inoculate them into other media for further experimentation.

Limitations

Bacteria that grow in liquid media may not have specific characteristics. Difficult to isolate different types of bacteria from mixed populations. Bacteria grow diffusely in liquids they produce discrete visible growth in solid media.

Safety considerations and standards

Sterilize all biohazardous waste before disposal. The use of a biological safety cabinet is recommended when culturing or processing specimens for fungi, mycobacteria, or for any procedure that may create dangerous aerosols. After use, all media, specimens, and containers must be sterilized by incineration or in an autoclave before disposal (121°C. for 30 min is recommended as a minimum). Care should be exercised in the opening of tubes with tight caps to prevent the breakage of the glass. Care should be taken to avoid contact with skin, eyes, or mucous membranes when handling culture media or any laboratory reagent, stain, fixative, or chemical. If contact occurs, flush immediately with running water. Contact a physician, hospital, or poison control center if overexposure or irritation exists.

Pour plate method for bacterial colony counting

Chapter outline

Principle: The pour plate method is a microbiological laboratory technique for iden-tifying and counting live bacteria in a liquid sample that is added with or before molten agar media before solidification. In general, this approach is used to count viable microorganisms in a particular sample by counting the total number of colony-forming units (CFUs) within and/or on the surface of the solid medium. It is mostly used to count bacteria, however Actinobacteria, moulds, and yeasts can also be isolated and counted.

Materials and equipment (materials and reagents)

- Mixed culture
- Inoculators
- Inoculating loop
- Sterile pipette
- Nutrient agar liquefied
- Sterile Petri dish

Step-by-step method details (experimental procedure)
Procedure
1. Label the bottom of the sterile Petri dish with the source of the culture and turn the plate.
2. Obtained two tubes of liquefied agar (nutrient agar) is boiled to melt the agar.
3. Agar at that temperature would kill the bacteria when are introduced so the agar is cooled to 60°C and are to be held in the water bath to maintain the temperature.
4. This would not kill the bacteria when they are introduced to the liquid agar and will reduce the amount of consideration that will get collected on the lid of the Petri dish.
5. The agar will solidify at 42°C after that the drop of culture is aseptically transferred to it.
6. Mix the tube by rolling it between your hands.
7. Pour the inoculated liquid into the sterile Petri dish that was labeled.
8. Now move the dish to cover the bottom of the disc with agar.
9. Allow the agar to solidify that would take an hour.
10. Incubate the plate at 37°C for 24—48 h.

Expected outcomes
The microbial colonies' growth was obtained on the surface of media as well as inside the media.

Advantages
Easy to undertake. Will detect lower concentrations than the surface spread method because of the larger sample volume. It requires no predrying of the agar surface. The most common method for determining the total viable count is the pour plate method. The pour plate technique can be used to determine the number of microbes/mL in a specimen. It has the advantage of not requiring previously prepared plates and is often used to assay bacterial contamination of foodstuffs.

Limitations
The use of relatively hot agar carries the risk of killing some sensitive contaminants, so giving a low result. Small colonies may be overlooked. In the case of solid sample dissolving in water, some species may suffer a degree of viability loss if diluted quickly in cold water; consequently, isotonic buffer (phosphate-buffered saline, for example) or peptone water are used as solvents or diluents. Colonies of different species within the agar appear similar—so it is difficult to detect contaminants. The reduced growth rate of obligate aerobes in the depth of the agar. Preparation for the pour plate method is time-consuming compared with streak plateand/or spread plate technique.

Safety considerations and standards

Decrease the risk of contamination by pouring plates in a laminar-airflow cabinet. When pouring multiple plates, flame the mouth of the flask before moving on to the next plate to reduce the risk of contamination. Keep the molten agar in the water bath for no more than 3–4 h. Don't remelt the agar. Use phosphate buffer pH 7.2 if necessary to dilute the suspension (Fig. 43.1).

Pipette inoculum onto sterile plate

Add sterile media

Swirl to mix and incubate

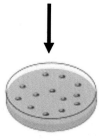

Colonies grow only on surface

FIGURE 43.1

Pour plate method.

Gram staining of bacteria

Chapter outline

Principle: Staining is a supporting technique used in microscopic techniques used to enhance the clarity of the microscopic image. Stains and dyes are widely used in the scientific field to highlight the structure of biological specimens, cells, tissues, etc. The most widely used staining procedure in microbiology is the gram staining, discovered by the Danish scientist and physician Hans Christian Joachim Gram in 1884. Gram staining is a differential staining technique that differentiates bacteria into two groups: gram-positive and gram-negative. The procedure is based on the ability of microorganisms to retain the color of the stains used during the gram stain reaction. The alcohol, losing the color of the primary stain, purple, decolorizes gram-negative bacteria. Gram-positive bacteria are not decolorized by alcohol and remain purple. After the decolorization step, a counter-stain is used to impart a pink color to the decolorized gram-negative organisms.

Materials and equipment (materials and reagents)

- Streaking plate
- Gram staining reagents:
- Crystal violet (as a primary stain and used as a histological stain)
- Gram's iodine solution (added as a mordant to enhance the crystal violet staining by forming crystal violet-iodine complex)
- 95% ethyl alcohol (acts as a decolorizing agent)

Basic Life Science Methods. https://doi.org/10.1016/B978-0-443-19174-9.00043-X

- Safranin (acts as a counter-stain)
- Wash bottle of distilled water
- Inoculating loop
- Glass slides
- Blotting paper
- Sprit lamp
- Microscope

Step-by-step method details (experimental procedure)

Procedure

1. Pick culture from the streaking plate with the help of a sterilized inoculation loop and drop it on a clean slide.
2. Drop a drop of water on the culture with a micropipette and spread in a circular motion.
3. Heat fixing the smear with the help of a spirit lamp.
4. Cover the smear with crystal violet for 1 min.
5. Wash the slide with distilled water for 30 s using a wash bottle.
6. Cover each smear with gram's iodine solution for 60 s.
7. Note: add ethyl alcohol drop by drop, until no more color flows from the smear.
8. Wash off iodine solution with 95% ethyl alcohol.
9. Wash the slides with distilled water and blot with blotting paper.
10. Apply safranin to smear for 30 s.
11. Again, wash with distilled water and blot with blotting paper.
12. Let the strained slide air dry.

Expected outcomes

The pink color was observed under the microscope, which confirms that the bacteria were gram-negative.

Advantages

It gives quick results when examining infections. It is simple and cost-effective. It helps with determining appropriate treatments for infection. It allows for various methods of testing.

It is basically a key procedure in identifying bacteria.

Limitations

1. Can't stain acid-fast bacilli (*Mycobacterium spp.*), and bacteria without cell wall like *Mycoplasma spp.* 2. Unsuitable for minute bacteria like *Rickettsia spp.*,

Chlamydia spp., etc. 3. Require multiple reagents. 4. Overdecolorization may result in the identification of false gram-negative results, whereas underdecolorization may result in the identification of false gram-positive results. 5. Smears that are too thick or viscous may retain too much primary stain, making the identification of proper gram stain reactions difficult. Gram-negative organisms may not decolorize properly. 6. Cultures older than 16—18 h will contain living and dead cells. Cells that are dead will be deteriorating and will not retain the stain properly. 7. The stain may form a precipitate with aging. Filtering through gauze will remove excess crystals. 8. Gram stains from patients on antibiotics or antimicrobial therapy may have altered Gram stain reactivity due to the successful treatment. 9. Occasionally, pneumococci identified in the lower respiratory tract on a direct smear will not grow in culture. Some strains are obligate anaerobes. 10. Toxin-producing organisms such as *Clostridia,* *Staphylococci,* and *Streptococci* may destroy white blood cells within a purulent specimen. 11. Faintly staining gram-negative organisms, such as *Campylobacter* and *Brucella,* may be visualized using an alternative counterstain (e.g., basic fuchsin).

Safety considerations and standards

Keep away from heat, hot surfaces, sparks, open flames, and other ignition sources. Avoid breathing dust/fume/gas/mist/vapors/spray. Avoid release to the environment. Wear protective gloves/protective clothing/eye protection/face protection. Rinse cautiously with water for several minutes. Remove contact lenses, if present and easy to do. Continue rinsing.

Casein hydrolysis test

Chapter outline

Principle: Casesase is an exoenzyme that is secreted out of the cells by some bacteria in order to degrade casein. The enzyme caseinase is an exoenzyme into the surrounding media, catalyzing the breakdown of milk protein, called casein, into small peptides and individual amino acids which are then taken up by the organism for energy use or as a building material. The hydrolysis reaction causes the milk agar, normally the opacity of real milk, to clear around the growth area as the casein protein is converted into soluble and transparent end products—small chains of amino acids, dipeptides, and polypeptides. The test determines whether an organism can produce the exoenzyme casesase or caseinase. Casein is a large protein that is responsible for the white color of milk. This test is conducted on milk agar which is a complex media containing casein, peptone, and beef extract, if an organism can produce casein, then there will be a zone of clearing around the growth.

Materials and equipment (materials and reagents)

- Casein 20 g
- Beef extract 1.5 g
- Peptone 5 g
- Agar 20 g

Basic Life Science Methods. https://doi.org/10.1016/B978-0-443-19174-9.00047-7

- NaCl 5g
- Distilled water 1000 mL

Step-by-step method details (experimental procedure)
Procedure

1. Prepare a nutrient media and autoclave it at 121°C 15 lbs pressure for 15 min.
2. Add 0.5 g casein powder after cooling the media.
3. Pour the media in Petri plate, wait till it solidify and streak with the culture to be tested on it.
4. Incubate plates at optimum temperature for 24 h at 37°C.
5. Observe the result after 24 h for presence or absence of clear zone around the streak.

Expected outcomes

A clear zone appeared around the streaking appears on the plate which represent positive casein test hydrolysis.

Advantages

Casein hydrolysis test helps us to identify the bacteria that is grown in the milk. It also helps us in differentiating the Enterobacteriaceae, Bacillaceae, and the other families. Casein hydrolysis test can also be used for differentiating the aerobic actinomycetes based on the proteolysis of casein.

Limitations

It is recommended that biochemical, immunological, molecular, or mass spectrometry testing be performed on colonies from pure culture for complete identification.

Safety considerations and standards

Observe universal precautions; wear protective gloves, laboratory coats, and safety glasses during all steps of this method. Discard any residual sample material by autoclaving after analysis is completed. Place disposable plastic, glass, and paper (pipet tips, autosampler vials, gloves, etc.) that contact plasma in a biohazard autoclave bag and keep these bags in appropriate containers until sealed and autoclaved. Wipe down all work surfaces with 10% bleach solution when work is finished. Handle acids and bases with extreme care; they are caustic and toxic. Handle organic solvents only in a well-ventilated area or, as required, under a chemical fume hood. For best results, a fasting sample should be obtained. Specimens for total homocysteine analysis may be fresh or frozen plasma.

Antibiotic resistance test by agar well diffusion method

46

Chapter outline

Principle: The performance of antimicrobial susceptibility testing is important to confirm susceptibility to choose empirical antimicrobial agents, or to detect resistance in individual bacterial isolates. The most widely used testing methods include the agar diffusion method and disc diffusion method.

Materials and equipment (materials and reagents)

- Broth culture of bacteria to be tested
- Antibiotics
- Nutrient agar plates
- Well cutter
- Bunsen burner
- Petri plates
- Conical flask
- Spreader

Step-by-step method details (experimental procedure)
Procedure
1. First weigh the amount of nutrient agar media to be required and add distilled water.
2. Then autoclave the media at 121°C, 15 lbs for 15 min.
3. After autoclaving place the media to cool up to 50°C, then pour it into the Petri dishes.
4. Keep the plates in Laminar Air Flow for a time period so that the media will solidify well.
5. After solidifying the media take a good cutter and make a well on the agar plates at a distance of 20 mm each.
6. After making wells spread the bacterial culture over the medium.
7. In these wells, pour the different drugs.
8. Do not invert the plates and place them in an incubator for the incubation period of 24 h at 35°C.
9. After the incubation period, observe the inhibition zone.

Expected outcomes
The zones of growth inhibition around each of the antibiotic discs are measured to the nearest millimeter. The diameter of the zone is related to the susceptibility of the isolate and to the isolate and to the diffusion rate of the drug through ate agar medium. The advantages of the agar diffusion method are the test simplicity that does not require any special equipment, the provision of categorical results easily interpreted by all clinicians, and flexibility in selection of drugs for testing. It is the least costly of all susceptibility methods (Table 46.1).

Table 46.1 The table below shows the result for the measured zone of inhibition for the respective antibiotics used.

Antibiotics	Zone of inhibition (mm)
Streptomycin	22
Oxytetracycline	20
Tetracycline	19
Ampicillin	NIL

Advantages

The main advantages are simplicity, reproducibility, ease in modifying antimicrobial discs, the possibility for use as a screening test against numerous isolates, and last but not least low-cost.

Limitations

Disadvantages are that some bacteria grow poorly or not at all on the media, and the minimum inhibitory concentration cannot be determined (Fig. 46.1).

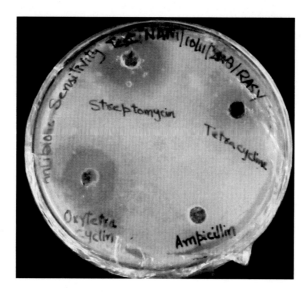

FIGURE 46.1

The above figure depicts that the isolated bacteria was resistant against the antibiotic Ampicillin as no zone of inhibition is being observed and was sensitive against the antibiotics streptomycin, oxytetracycline, and tetracycline.

Index

Printed in the United States
by Baker & Taylor Publisher Services